CW00409541

TIME SERIES ANALYSIS AND INVERSE THEORY FOR GEOPHYSICISTS

Geophysicists make measurements at the Earth's surface, and from aeroplanes and satellites, in order to infer structure inside the Earth. The digital revolution now provides us with vast datasets that are interpreted by using sophisticated processing methods. This unique textbook provides the foundation for understanding and applying those techniques commonly used in geophysics to process and interpret modern digital data. The digital revolution is only a decade old, but it has changed all aspects of geophysical field measurement. Sophisticated methods are needed to maximise the information contained within a modern dataset, and these make heavy demands on the mathematical expertise of the modern geophysics student.

The data analyst's toolkit contains a wide range of techniques that may be divided into two main groups: processing, which concerns mainly time series analysis and is used to separate the signal of interest from background noise; and inversion, which involves generating some map or physical model from the data. These two groups of techniques are normally taught separately, but are here presented together as Parts I and II of the book. Part III describes some real applications taken from the author's experience in geophysical research. They include case studies in seismology, geomagnetism and gravity.

This textbook gives students and practitioners the theoretical background and practical experience, through case studies, computer examples and exercises, to understand and apply new processing methods to modern geophysical datasets. Suitable for undergraduate- and graduate-level courses, these methods are equally applicable to other disciplines. Files needed for the computer exercises are available on a website at http://publishing.cambridge.org/resources/0521819652. Solutions to the exercises are also available to tutors through this same website.

DAVID GUBBINS was awarded a Ph.D. from the Department of Geodesy and Geophysics, University of Cambridge in 1972. After several years spent as a researcher in the University of Colorado, the Massachusetts Institute of Technology and the University of California at Los Angeles, he returned to Cambridge University where he held the position of Assistant Director of Research until 1989. In 1989 he joined the School of Earth Sciences in the University of Leeds where he is now Professor of Geophysics. Professor Gubbins is a member of the Society of Exploration Seismologists and the Royal Astronomical Society, and has been elected a Fellow of the American Geophysical Union, of the Royal Society of London, and of the Institute of Physics. His contributions to geophysics are mainly in geomagnetism and seismology. His work has been recognised through numerous prestigious awards and visiting lectureships including the Murchison Medal of the Geological Society of London and the Gold Medal of the Royal Astronomical Society. Professor Gubbins has taught mathematics courses to science students for over 35 years and is also the author of the acclaimed textbook *Seismology and Plate Tectonics* (Cambridge University Press, 1990).

TIME SERIES ANALYSIS AND INVERSE THEORY FOR GEOPHYSICISTS

DAVID GUBBINS
Department of Earth Sciences,
University of Leeds

CAMBRIDGE
UNIVERSITY PRESS

PUBLISHED BY THE PRESS SYNDICATE OF THE UNIVERSITY OF CAMBRIDGE
The Pitt Building, Trumpington Street, Cambridge, United Kingdom

CAMBRIDGE UNIVERSITY PRESS
The Edinburgh Building, Cambridge CB2 2RU, UK
40 West 20th Street, New York, NY 10011–4211, USA
477 Williamstown Road, Port Melbourne, VIC 3207, Australia
Ruiz de Alarcón 13, 28014 Madrid, Spain
Dock House, The Waterfront, Cape Town 8001, South Africa

http://www.cambridge.org

First published 2004

Printed in the United Kingdom at the University Press, Cambridge

Typeface Times 11/14 pt. *System* LaTeX 2_ε [TB]

A catalogue record for this book is available from the British Library

Library of Congress Cataloguing in Publication data

Gubbins, David.
Time series analysis and inverse theory for geophysicists / David Gubbins.
p. cm.
Includes bibliographical references and index.
ISBN 0 521 81965 2 – ISBN 0 521 52569 1 (paperback)
1. Earth science – Mathematics. 2. Time-series analysis. 3. Inversion (Geophysics) I. Title.
QC809.M37G83 2004
550′.1′51 – dc22 2003055730

ISBN 0 521 81965 2 hardback
ISBN 0 521 52569 1 paperback

Dedication

To my teachers, Rodney Spratley, John Hills, Brian Pippard, and to the memory of Teddy Bullard

Contents

	Preface	*page* xi
	Acknowledgements	xiii
	List of illustrations	xiv
1	Introduction	1
	1.1 The digital revolution	1
	1.2 Digital recording	3
	1.3 Processing	5
	1.4 Inversion	7
	1.5 About this book	10
	Exercises	12
Part I	**Processing**	15
2	Mathematical preliminaries: the z- and discrete Fourier transforms	17
	2.1 The z-transform	17
	2.2 The discrete Fourier transform	21
	2.3 Properties of the discrete Fourier transform	26
	2.4 DFT of random sequences	34
	Exercises	36
3	Practical estimation of spectra	40
	3.1 Aliasing	40
	3.2 Spectral leakage and tapering	44
	3.3 Examples of spectra	49
	Exercises	53
4	Processing of time sequences	57
	4.1 Filtering	57
	4.2 Correlation	63
	4.3 Deconvolution	65
	Exercises	69

5 Processing two-dimensional data 73
- 5.1 The 2D Fourier transform 73
- 5.2 2D filtering 75
- 5.3 Travelling waves 77
- Exercises 80

Part II Inversion 83

6 Linear parameter estimation 85
- 6.1 The linear problem 85
- 6.2 Least-squares solution of over-determined problems 89
- 6.3 Weighting the data 92
- 6.4 Model covariance matrix and the error ellipsoid 100
- 6.5 Robust methods 103
- Exercises 107

7 The under-determined problem 110
- 7.1 The null space 110
- 7.2 The minimum-norm solution 112
- 7.3 Ranking and winnowing 113
- 7.4 Damping and the trade-off curve 115
- 7.5 Parameter covariance matrix 117
- 7.6 The resolution matrix 121
- Exercises 123

8 Nonlinear inverse problems 125
- 8.1 Methods available for nonlinear problems 125
- 8.2 Earthquake location: an example of nonlinear parameter estimation 127
- 8.3 Quasi-linearisation and iteration for the general problem 130
- 8.4 Damping, step-length damping, and covariance and resolution matrices 131
- 8.5 The error surface 132
- Exercises 135

9 Continuous inverse theory 138
- 9.1 A linear continuous inverse problem 138
- 9.2 The Dirichlet condition 139
- 9.3 Spread, error and the trade-off curve 142
- 9.4 Designing the averaging function 144
- 9.5 Minimum-norm solution 145
- 9.6 Discretising the continuous inverse problem 147
- 9.7 Parameter estimation: the methods of Backus and Parker 149
- Exercises 154

Part III Applications 157
10 Fourier analysis as an inverse problem 159
10.1 The discrete Fourier transform and filtering 159
10.2 Wiener filters 161
10.3 Multi-taper spectral analysis 164
11 Seismic travel times and tomography 170
11.1 Beamforming 170
11.2 Tomography 177
11.3 Simultaneous inversion for structure and earthquake location 183
12 Geomagnetism 188
12.1 Introduction 188
12.2 The forward problem 189
12.3 The inverse problem: uniqueness 190
12.4 Damping 193
12.5 The data 196
12.6 Solutions along the trade-off curve 199
12.7 Covariance, resolution and averaging functions 203
12.8 Finding fluid motion in the core 207
Appendix 1 Fourier series 213
Appendix 2 The Fourier integral transform 218
Appendix 3 Shannon's sampling theorem 224
Appendix 4 Linear algebra 226
Appendix 5 Vector spaces and the function space 234
Appendix 6 Lagrange multipliers and penalty parameters 241
Appendix 7 Files for the computer exercises 244
References 246
Index 250

Preface

The digital revolution has replaced traditional recording in the form of graphs on paper with numbers written to some form of magnetic or optical recording. Digital data can be processed to emphasise some aspect of the signal, an enormous advantage over paper records. For the student, the digital revolution has meant learning a whole host of new techniques, most of them based on quite advanced mathematics. The main purpose of this book is to provide the student of geophysics with an introduction to these techniques and an understanding of the underlying philosophy and mathematical theory.

The book is based on two courses taught to Bachelors and Masters students at Leeds over the past 10 years, one on Time Series in the second undergraduate year and one on Inversion in the third. The 3-year degree programme in the UK presents a problem: the techniques must be learnt in the second year if they are to be applied in the third. Time series analysis relies heavily on Fourier analysis, and although second year students have met Fourier series they have not met the Fourier integral theorem. This book makes a virtue of necessity by avoiding the Fourier integral transform and using only the discrete transform, for which we only need the sum of a geometrical series. I have come to see this as an advantage because modern data come in a discrete form, rather than as continuous functions, and are finite in duration, rather than going on forever.

The second problem arose with inverse theory, a notoriously difficult subject that owes much of its development to geophysicists such as George Backus, Freeman Gilbert, Bob Parker, Albert Tarantola, and others too numerous to mention. The subject is hard because it uses functional analysis and the theory of Hilbert spaces, which are beyond the scope of most mathematics courses taught to scientists. Once again, the subject is greatly simplified by a discrete treatment, which reduces the mathematics to matrix algebra. I have tried to do this without losing the essential flavour of inversion proper, and by distinguishing inversion from simple parameter estimation. Chapter 9, on continuous inversion, is an afterthought. This material has never been taught, but I could not resist an attempt to Bowdlerise the elegant Backus and Parker theories.

Acknowledgements

I owe a debt of thanks to all my colleagues at Leeds who have helped and advised over the development period of these courses. Jurgen Neuberg was instrumental in setting the style for the time series classes, while Roger Clark, Derek Fairhead, Greg Houseman, Andy Jackson and Graham Stuart have provided most welcome input at one time or another. George Helffrich and Richard Holme read and provided comments on early drafts. Most of all I am indebted to the generations of Leeds students who have struggled with the difficult material and provided feedback to improve the courses, and to those research students who helped teach the practicals. Of course, I take all responsibility for any errors that remain in the book.

Illustrations

1.1	Seismogram of regional earthquake	*page* 05
1.2	First arrival detail of regional earthquake	06
1.3	Gravity anomaly	08
2.1	The convolution process	20
2.2	The unit circle	22
2.3	Boxcar	25
2.4	Cyclic convolution	29
2.5	Stromboli velocity/displacement	31
3.1	Aliasing in the time domain	41
3.2	Aliasing in the frequency domain	42
3.3	Normal modes for different window lengths, boxcar window	45
3.4	Tapers	48
3.5	Proton magnetometer data	50
3.6	Airgun source depth	52
3.7	Microseismic noise	53
4.1	Aftershock for Nov. 1 event	58
4.2	Butterworth filters, amplitude spectra	59
4.3	Low-pass filtered seismograms: effect of n	60
4.4	Causal and zero-phase filters compared	61
4.5	Water level method	66
5.1	2D filtering	74
5.2	2D aliasing	75
5.3	Upward continuation	78
5.4	Separation by phase speed	79
5.5	Example of f–k filtering	79
6.1	Two-layer density model	87
6.2	Picking errors	93
6.3	Dumont D'Urville jerk	96

6.4	Error ellipse	101
6.5	Moment of inertia and mass covariance	103
6.6	L^1 vs L^2 norm results	106
7.1	Null space for mass/inertia problem	111
7.2	Trade-off curve	116
8.1	Hypocentre partial derivatives	128
8.2	Error surfaces	133
9.1	Data kernels K_1, K_2 for mass and moment of inertia	139
9.2	Averaging function for mass of Earth's core	145
10.1	Optimal tapers	167
11.1	Stacking of seismograms from borehole data	174
11.2	Stacking of seismograms from regional earthquake	175
11.3	Filtered seismograms from regional earthquake	176
11.4	1D tomography	178
11.5	ACH inversion	181
12.1	Data kernels for main geomagnetic field	194
12.2	Data distribution plots	198
12.3	Trade-off curve	201
12.4	Solutions for epoch 1980	202
12.5	Contour maps of errors	204
12.6	Resolution matrices	206
12.7	Averaging functions	208
A1.1	Square wave	217

1
Introduction

1.1 The digital revolution

Recording the output from geophysical instruments has undergone four stages of development during the past century: mechanical, optical, analogue magnetic, and digital. Take the seismometer as a typical example. The principle of the basic sensor remains the same: the swing of a test mass in response to motion of its fixed pivot is monitored and converted to an estimate of the velocity of the pivot.

Inertia and damping determine the response of the sensor to different frequencies of ground motion; different mechanical devices measured different frequency ranges. Ocean waves generate a great deal of noise in the range 0.1–0.5 Hz, the *microseismic noise* band, and it became normal practice to install a short-period instrument to record frequencies above 0.5 Hz and a long-period instrument to record frequencies below 0.1 Hz.

Early mechanical systems used levers to amplify the motion of the mass to drive a pen. The classic short-period, high-gain design used an inverted pendulum to measure the horizontal component of motion. A large mass was required simply to overcome friction in the pen and lever system.

An optical lever reduces the friction dramatically. A light beam is directed onto a mirror, which is twisted by the response of the sensor. The reflected light beam shines onto photographic film. The sensor response deflects the light beam and the motion is recorded on film. The amplification is determined by the distance between the mirror and film. Optical recording is also compact: the film may be enlarged later to a more readable size. Optical recording was in common use in the 1960s and 1970s.

Electromechanical devices allow motion of the mass to be converted to a voltage, which is easy to transmit, amplify and record. Electromagnetic feedback seismometers use a null method, in which an electromagnet maintains the mass in a constant position. The voltage required to drive the electromagnet is monitored and forms the output of the sensor.

This voltage can be recorded on a simple tape recorder in analogue form. Analogue magnetic records could be converted to paper records simply by playing them through a chart recorder. There is a good analogy with tape recording sound, since seismic waves are physically very similar to low-frequency sound waves. The concept of fidelity of recording carries straight across to seismic recording. A convenient way to search an analogue tape for seismic sources is to simply play it back fast, thus increasing the frequency into the audio range, and listen for bangs.

The digital revolution started in seismology in about 1975, notably when the World Wide Standardized Seismograph Network (WWSSN) was replaced by Seismological Research Observatories (SROs). These were very expensive installations requiring a computer in a building on site. The voltage is sampled in time and converted to a number for input to the computer. The tape recording systems were not able to record the incoming data all the time so the instrument was triggered and a short record retained for each event. Two channels (sometimes three) were output: a finely sampled short-period record for the high-frequency arrivals and a coarsely sampled channel (usually one sample each second) for the longer-period surface waves. Limitations of the recording system meant that SROs did not herald the great revolution in seismology some had hoped for: that had to wait for better mass storage devices.

The great advantage of digital recording is that it allows replotting and processing of the data after recording. If a feature is too small to be seen on the plot, you simply plot it on a larger scale. More sophisticated methods of processing allow us to remove all the energy in the microseismic noise band, obviating the need for separate short- and long-period instruments. It is even possible to simulate an older seismometer simply by processing, provided the sensor records all the information that would have been captured by the simulated instrument. This is sometimes useful when comparing seismograms from different instruments used to record similar earthquakes. Current practice is therefore to record as much of the signal as possible and process it after recording. This has one drawback: storage of an enormous volume of data.

The storage problem was essentially solved in about 1990 by the advent of cheap hard disks and tapes with capacities of several gigabytes. Portable broadband seismometers were developed at about the same time, creating a revolution in digital seismology: prior to 1990, high-quality digital data were available only from a few permanent, manned observatories. After 1990 it was possible to deploy arrays of instruments in temporary sites to study specific problems, with only infrequent visits to change disks or tapes.

1.2 Digital recording

The *sensor* is the electromechanical device that converts physical input (e.g. ground motion) into voltage; the *recorder* converts the voltage into numbers and stores them. The ideal sensor would produce a voltage that is proportional to the ground motion but such a device is impossible to make (the instrument response would have to be constant for all frequencies, which requires the instrument to respond instantaneously to any input, see Appendix 2). The next best thing is a *linear response*, in which the output is a convolution of the ground motion with the *transfer function* of the instrument.

Let the voltage output be $a(t)$. The recorder *samples* this function regularly in time, at a *sampling interval* Δt, and creates a sequence of numbers:

$$\{a_k = a(k\Delta t); \quad k = 0, 1, 2, \ldots, N\}. \tag{1.1}$$

The recorder stores the number as a string of bits in the same way as any computer.

Three quantities describe the limitations of the sensor: the *sensitivity* is the smallest signal that produces non-zero output; the *resolution* is the smallest change in the signal that produces non-zero output; and the *linearity* determines the extent to which the signal can be recovered from the output. For example, the ground motion may be so large that the signal exceeds the maximum level; the record is said to be 'clipped'. The recorded motion is not linearly related to the actual ground motion, which is lost.

The same three quantities can be defined for the recorder. A pen recorder's linearity is between the voltage and movement of the pen, which depends on the electronic circuits and mechanical linkages; its resolution and accuracy are limited by the thickness of the line the pen draws. For a digital recorder, linearity requires faithful conversion of the analogue voltage to a digital count, while resolution is set by the voltage corresponding to one digital count.

The recorder suffers two further limitations: the *dynamic range*, g, the ratio of maximum possible to minimum possible recorded signal, usually expressed in deciBels: $20\log_{10}(g)$ dB; and the maximum frequency that can be recorded. For a pen recorder the dynamic range is set by the height of the paper, while the maximum frequency is set by the drum speed and width of the pen. For a digital recorder the dynamic range is set by the number of bits available to store each member of the time sequence, while the maximum frequency is set by the sampling interval Δt or *sampling frequency* $\nu_s = 1/\Delta t$ (we shall see later that the maximum meaningful frequency is in fact only half the sampling frequency).

A recorder employed for some of the examples in this book used 16 bits to record the signal as an integer, making the maximum number it can record $2^{16} - 1$ and the minimum is 1. The dynamic range is therefore $20\log_{10}(65\,535) = 96$ dB.

Another popular recorder uses 16 bits but in a slightly more sophisticated way. One bit is used for the sign and 15 are used for the integer if the signal is in the range $\pm 2^{15} - 1$; outside this range it steals one bit from the integer, records an integer in the range $\pm(2^{14} - 1)$, and multiplies by 20, giving a complete range $\pm 327\,680$, or 116 dB. The stolen 'gain' bit is used to indicate the change in gain by a factor of 20; the increase in dynamic range has been achieved at the expense of accuracy, but this is usually unimportant because it occurs only when the signal is large. A typical 'seismic word' for recorders in exploration geophysics consists of 16 bits for the integer ('mantissa'), one bit for the sign, and four bits for successive levels of gain. The technology is changing rapidly and word lengths are increasing; most recorders now being sold have 24 bits.

Storage capacity sets a limit to the dynamic range and sampling frequency. A larger dynamic range requires a larger number and more bits to be stored; a higher sampling frequency requires more numbers per second and therefore more numbers to be stored. The following calculation gives an idea of the logistical considerations involved in running an array of seismometers in the field. In 1990, Leeds University Earth Sciences Department deployed nine three-component broadband instruments in the Tararua Mountain region of North Island, New Zealand, to study body waves travelling through the subducted Pacific Plate. The waves were known to be high frequency, demanding a 50 Hz sampling rate. Array processing (see Section 11.1) is only possible if the signal is coherent across the array, requiring an interval of 10 km between stations. The Reftek recorders used a 16 bit word and had 360 Mb disks that were downloaded onto tape when full.

A single field seismologist was available for routine servicing of the array, which meant he or she had to drive around all nine instruments at regular intervals to download the disks before they filled. How often would they need to be downloaded? Each recorder was storing 16 bits for each of three components 50 times each second, or 2400 bit s^{-1}. Allowing 20% overhead for things like the time channel and state-of-health messages gives 2880 bit s^{-1}. One byte is eight bits and, dividing 2880 into the storage capacity of the disk, a 360 Mb disk can store $8 \times (3.6\ 10^8)$ bits, which would last about 10^6 s, or 11.5 days. It would be prudent to visit each instrument at least every 10 days, which is possible for an array stretching up to 150 km from base over good roads. A later deployment used Mars recorders with optical disks that could be changed by a local unskilled operator, requiring only infrequent visits to collect data and replenish the stock of blank disks. In this case, the limiting factor was the time required to back up the optical disks onto tape.

It is well worth making the effort to capture all the information available in a broadband seismogram because the information content is so extraordinarily rich. The seismogram shown in Figures 1.1 and 1.2 provides a good example. *P*, *S*, and

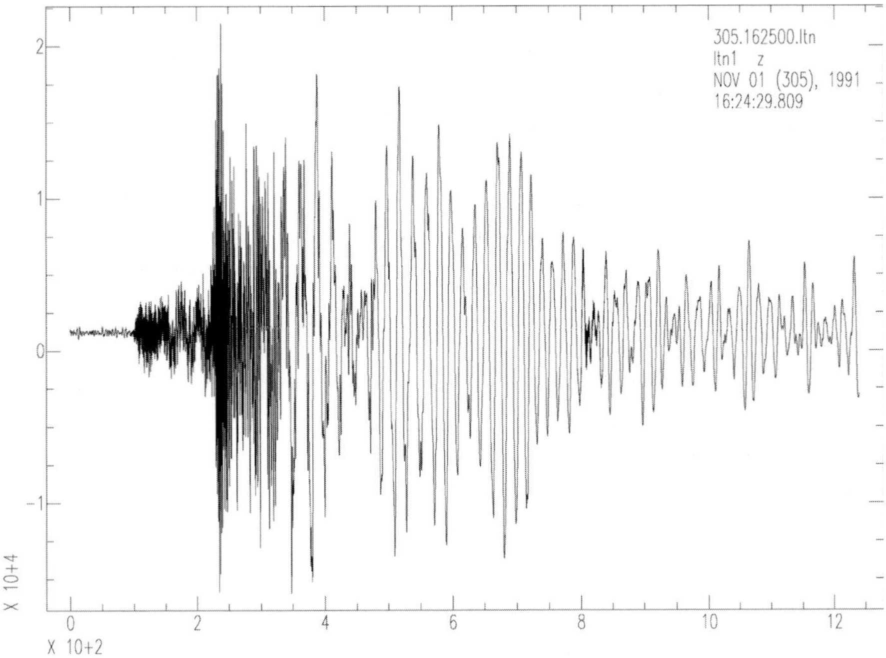

Fig. 1.1. Seismogram of an event in the Kermadec Islands on 1 November 1991 recorded in North Island, New Zealand. Note the *P*, *S*, and longer-period surface waves. The dynamic range meant maximum signals of ±16 383 digital counts, and the scale shows that this was nearly exceeded. The time scale is in minutes, and the surface waves have a dominant period of about 20 s.

surface waves are clearly seen. The surface waves are almost clipped (Figure 1.1), yet the onset of the *P* wave has an amplitude of just one digital count (Figure 1.2). The frequency of the surface waves is about 20 s, yet frequencies of 10 Hz and above may be seen in the body waves. The full dynamic range and frequency bandwidth were therefore needed to record all the ground motion.

1.3 Processing

Suppose now that we have collected some important data. What are we going to do with them? Every geophysicist should know that the basic raw data, plus that field note-book, constitute the maximum factual information he or she will ever have. This is what we learn on field courses. Data processing is about extracting a few nuggets from this dataset; it involves changing the original numbers, which *always* means losing information. So we always keep the original data.

Processing involves operating on data in order to isolate a *signal*, the message we are interested in, and to separate it from 'noise' , which nobody is interested in,

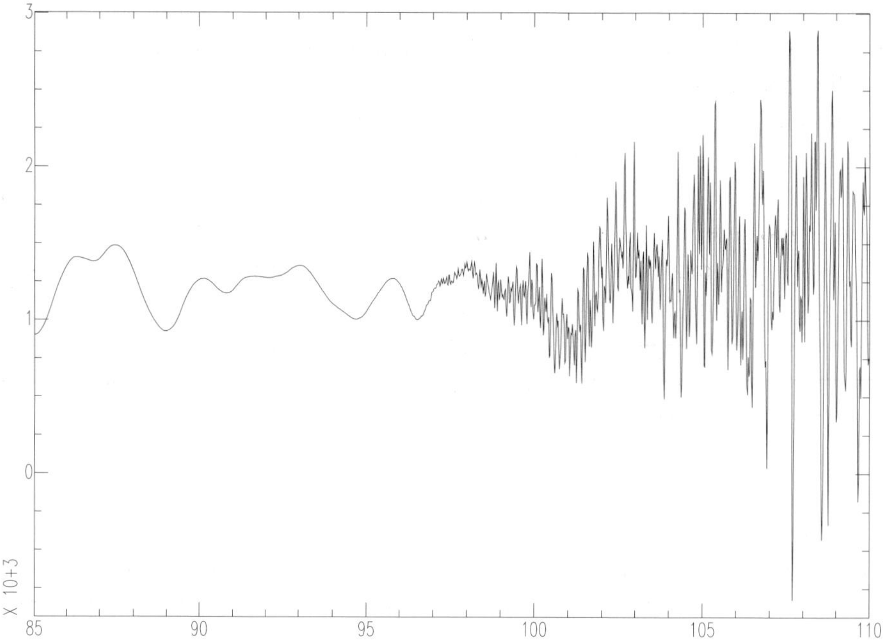

Fig. 1.2. Expanded plot of the first part of the seismogram in Figure 1.1. Note the very small amplitude of the *P* arrival, which rides on microseisms of longer period. The arrival is easy to see because its frequency is different from that of the microseisms.

and unwanted signals, which do not interest us at this particular time. For example, we may wish to examine the *P* wave on a seismogram but not the surface waves, or we may wish to remove the Earth's main magnetic field from a magnetic measurement in order to determine the magnetisation of local rocks because it will help us understand the regional geology. On another day we may wish to remove the local magnetic anomalies in order to determine the main magnetic field because we want to understand what is going on in the Earth's core.

Often we do not have a firm idea of what we ought to find, or exactly where we should find it. Under these circumstances it is desirable to keep the methods as flexible as possible, and a very important part of modern processing is the interaction between the interpreter and the computer. The graphical display is a vital aid in interpretation, and often the first thing we do is plot the data in some way. For example, we may process a seismogram to enhance a particular arrival by filtering (Section 4.1). We know the arrival time roughly, but not its exact time. In fact, our main aim is to measure the arrival time as precisely as possible; in practice, the arrival will be visible on some unprocessed seismograms, on others it will be visible only after processing, while on yet more it may never be visible at all.

The first half of this book deals with the analysis of time series and sequences, the commonest technique for processing data interactively without strict prior prejudice about the detailed cause of the signals or the noise. This book adopts Claerbout's 1992 usage and restricts the word *processing* to distinguish this activity from the more formal process of *inversion* explained below. Processing is interactive and flexible; we are looking for something interesting in the data without being too sure of what might be there.

Suppose again that we wish to separate P waves from surface waves. We plot the seismogram and find the P waves arriving earlier because they travel faster than surface waves, so we just cut out the later part of the seismogram. This does not work when the P waves have been delayed (by reflecting from some distant interface, for example) and arrive at the same time as the surface waves. The surface waves are much bigger than body waves and the P wave is probably completely lost in the raw seismogram.

P waves have higher frequencies than surface waves (period 1 s against 20 s for earthquake seismograms, or 0.1 s against 1 s for a typical seismic exploration experiment). This gives us an alternative means of separation. The Fourier transform, in its various forms (Appendix 2), decomposes a time series into its component frequencies. We calculate and plot the transform, identify the low-frequency contributions from the surface waves, zero them and transform back to leave the higher-frequency waves. This process is called *filtering* (Section 4.1). In these examples we need only look at the seismogram or its Fourier transform to see two separate signals; having identified them visually it is an easy matter to separate them. This is processing.

1.4 Inversion

The processed data are rarely the final end product: some further interpretation or calculation is needed. Usually we will need to convert the processed data into other quantities more closely related to the physical properties of the target. We might want to measure the arrival time of a P wave to determine the depth of a reflector, then interpret that reflector in the context of a hydrocarbon reservoir. We might measure spatial variations in the Earth's gravity, but we really want to find the density anomalies that cause those gravity anomalies, then understand the hidden geological structure that caused the gravity variations.

Inversion is a way of transforming the data into more easily interpretable physical quantities: in the example above we want to invert the gravity variations for density. Unlike processing, inversion is a formal, rigid procedure. We have already decided what is causing the gravity variations, and probably have an idea of the depth, extent, and even the shape, of the gravity anomalies. We have a

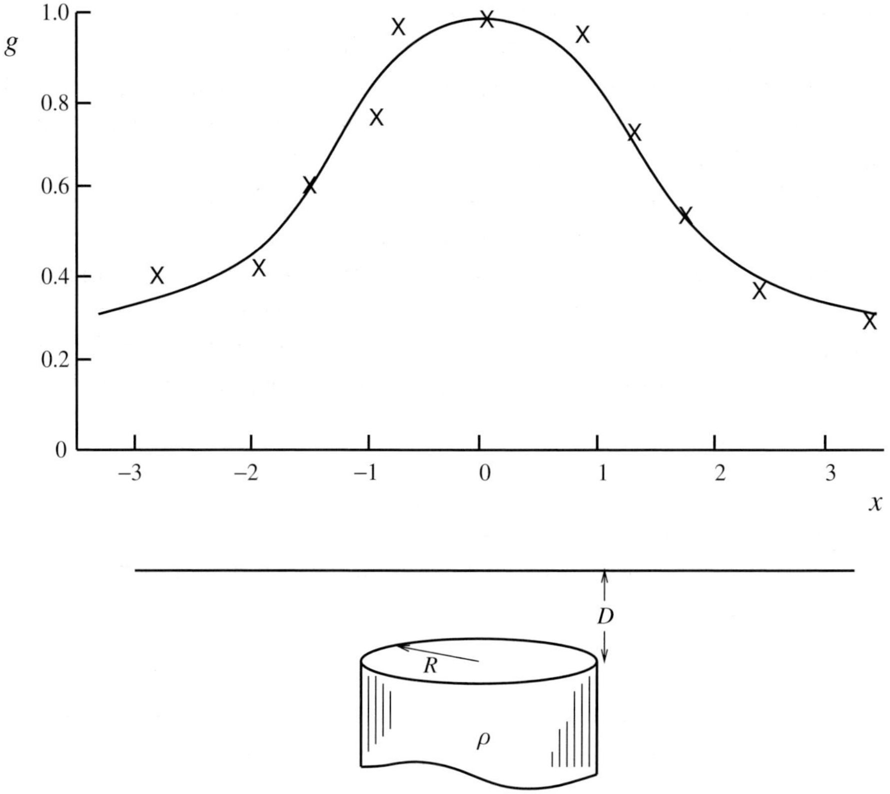

Fig. 1.3. Gravity anomaly caused by a dense, buried cylinder.

mathematical *model* that we wish to test and refine using this new data. The inversion excludes any radically different interpretation. For example, if we invert the seismic arrival times for the depth of a pair of horizontal reflectors we would never discover that they really come from a single, dipping reflector.

Consider the gravity traverse illustrated in Figure1.3, which we intend to invert for density. The traditional method, before widespread use of computers, was to compare the shape of the anomaly with theoretical curves computed from a range of plausible density models simple enough for the gravity to be calculated analytically. This is called *forward modelling*. It involves using the laws of physics to predict the observations from a model. Computers can calculate gravity signals from very complicated density models. They can also be used to search a large set of models to find those that fit the data. A strategy or algorithm is needed to direct the search. Ideally we should start with the most plausible models and refine our ideas of plausibility as the search proceeds. These techniques are sometimes called *Monte Carlo* methods.

There are two separate problems, *existence* and *uniqueness*. The first requirement is to find one model, any model, that fits the data. If none is found, the data are incompatible with the model. This is rare. Either there is something wrong with the model or we have overestimated the accuracy of the data. Having found one model we search for others. If other models are found the solution is nonunique; we must take account of it in any further interpretation. This always happens. The nonuniqueness is described by the subset of models that fits the data.

It is more efficient, when possible, to solve directly for the model from the data. This is not just an exercise in solving the equations relating the data to the model for a model solution, we must also characterise the nonuniqueness by finding the complete set of compatible solutions and placing probabilities on the correctness of each individual model.

Like most formal procedures, the mathematics of inverse theory is very attractive: it is easy to become seduced into thinking that the process is more important than it really is. Throughout this book I try to emphasise the importance of the measured data and the desired end goal: these are much more important than the theory, which is just the vehicle that allows you to proceed from the data to a meaningful interpretation. It is like a car that lets you take luggage to a destination: you must pack the right luggage and get to the right place (and, incidentally, the place should be interesting and worthy of a visit!). You need to know enough about the car to drive it safely and get there in reasonable time, but detailed knowledge of its workings can be left to the mechanic.

There are two sources of error that contribute to the final solution: one arising because the original measurements contain errors, and one arising because the measurements failed to sample some part of the model. Suppose we can prove for a particular problem that perfect, error-free data would invert to a single model: then in the real case of imperfect data we need only worry about measurement errors mapping into the model. In this book I call this *parameter estimation* to distinguish it from true inversion.

Modern inversion deals with the more general case when perfect data fail to provide a unique model. True inversion is often confused with parameter estimation in the literature. The distinction is vital in geophysics, because the largest source of error in the model usually comes from failure to obtain enough of the right sort of data, rather than sufficiently accurate data.

The distinction between inversion and parameter estimation should not become blurred. It is tempting to restate the inverse problem we should be solving by restricting the model until the available data are capable of determining a unique solution in the absence of errors, but this is philosophically wrong. The model is predetermined by our knowledge of the problem at the start of the experiment, not by what we are about to do.

Some quotes illustrate the diversity of meaning attached to the term 'inversion'. J. Claerbout, in his book *Earth Soundings Analysis: Processing Versus Inversion*, calls it matrix inversion. This is true only if the forward problem is posed as matrix multiplication. In exploration geophysics one sometimes hears inversion called *deconvolution* (Section 4.3), an even more restrictive definition. Deconvolution is treated as an inverse problem in Section 10.2. 'Given the solution of a differential equation, find its coefficients' was the definition used by Gel'fand and Levitan (1955). More specifically 'given the eigenvalues of a differential operator, find the operator', a problem that finds application in the use of normal mode frequencies in determining average Earth structure. A more poetic statement of the same problem is 'can we hear the shape of a drum?', which has received much attention from pure mathematicians. None of these definitions covers the full extent of inverse problems currently being studied in geophysics.

1.5 About this book

The book is divided into three parts: processing, inversion, and applications that combine techniques introduced in both of the first parts. Emphasis is placed on discrete, rather than continuous, formulations, and on deterministic, rather than random, signals. This departure from most texts on signal processing and inversion demands some preliminary explanation.

The digital revolution has made it easy to justify treating data as a discrete sequence of numbers. Analogue instruments that produce continuous output in the form of a paper chart record might have the appearance of continuity, but they never had perfect time resolution and are equivalent to discrete recordings with interpolation imposed by the nature of the instrument itself – an interpolation that is all too often beyond the control and sometimes even the knowledge of the operator. The impression of continuity is an illusion. Part I therefore uses the discrete form of the Fourier transform, which does not require prior knowledge of the integral form.

It is much harder to justify discretising the model in an inversion. In this book I try to incorporate the philosophy of continuous inversion within the discrete formulation and clarify the distinction between true inversion, in which we can only ever discover a part of the true solution, and parameter estimation, where the model is adequately described by a few specially chosen numbers. Chapter 9 contains a brief overview of continuous inversion.

In the example of Figure 1.3 the model is defined mathematically by a set of parameters: R, the radius of the cylinder, D, the depth of burial, and ρ, the density or density difference with the surroundings. The corresponding gravity field is computed using Newton's inverse square law for each mass element $d\rho$ and integrating over the volume. This gives a prediction for every measured value. We can either select a curve and find the best fit by trial and error, then deduce the best values of the three parameters, or do a full mathematical inversion using the methods described in the Part II. Either way should get the same answer, but both will suffer from the same obvious drawback: the model is implausibly over-simplified. It is a pity to use the power of computers simply to speed up old procedures. It is better to prescribe a more general model and determine as much about it as the data allow. Ultimately, for this problem, this might mean allowing the density to vary with all three space coordinates, an infinite number of degrees of freedom.

For the most part, we shall content ourselves with a finite number of model parameters, but still allow the model to be as complicated as we wish, for example by dividing the half space into rectangular blocks and ascribing a different density to each cell. Some such *parametrisation* will be needed ultimately to represent the model numerically whatever inversion procedure is adopted. The methods are sufficiently flexible to allow incorporation of any previous knowledge we might have about the structure without necessarily building it into the parametrisation.

In practice, virtually all real inverse problems end up on the computer, where any continuous function is represented discretely. Continuous inversion has one great pedagogical advantage: if the model is a continuous function of an independent variable and therefore contains an infinite number of unknowns, it is difficult to pretend that a finite number of data can provide all the answers. Given a discrete formulation and a huge amount of (very expensive) data, it becomes all too easy to exaggerate the results of an inversion. Chapter 9 gives a brief introduction to the techniques of continuous inversion. Its main purpose is to explain how continuous inversion fits with discrete inversion, and also to illustrate some potential pitfalls of discrete treatments, rather than providing a practical approach to solving continuous inverse problems.

Deterministic signals are predictable. An error-free seismogram can be computed from the properties of the source and the medium the waves pass through; a gravity traverse can be computed from density; and a magnetometer survey can be predicted from magnetisation. For random signals we can only predict statistical properties: the mean, standard deviation, etc. Wind noise on a seismogram is one example, its statistical properties depending on the weather, neighbouring trees, etc. Electronic noise is another source of error in all modern digital instruments.

Many books use a statistical definition of the power spectrum (see, for example, Percival and Walden (1998); Priestley (1992)) whereas this book uses the more familiar mathematical definition. The central problem in transferring the familiar mathematical techniques of Fourier series and transforms to data analysis is that both assume the time function continues forever: Fourier series require periodicity in time while the Fourier integral transform (FIT)(Appendix 2) requires knowledge for all time. Data series never provide this, so some assumptions have to be made about the signal outside the measurement range. The assumptions are different for random and deterministic signals. For example, it would seem reasonable to assume a seismogram is zero outside the range of measurement; the same could apply to a gravity traverse if we had continued right across the anomaly. Random signals go on for ever yet we only measure a finite portion of them. They are often continued by assuming *stationarity*, that the statistical properties do not vary with time.

The emphasis on deterministic signals in this book is prompted by the subject. The maxim 'one man's signal is another man's noise' seems to apply with more force in geophysics than in other branches of physical science and engineering. For example, our knowledge of the radius of the Earth's core is limited not by the accuracy with which we can read the arrival time of seismic reflections from the boundary, but from the time delays the wave has acquired in passing through unknown structures in the mantle. Again, the arrival times of shear waves are made inaccurate because they appear within the P wave coda, rather than at a time of greater background noise. The main source of error in magnetic observations of the Earth's main field, the part that originates in the Earth's core, comes from magnetised rocks in the crust, not from inherent measurement errors – even for measurements made as early as the late eighteenth century. One has to go back before Captain Cook's time, when navigation was poor, to find measurement errors comparable with the crustal signal.

Something similar applies to errors in models obtained from an inversion. Gravity data may be inverted for density, but an inherent ambiguity makes it impossible to distinguish between small, shallow-mass anomalies from large, deep ones. Seismic tomography (Chapter 11), in which arrival times of seismic waves are inverted for wave speeds within the Earth, is limited more by its inability to distinguish anomalies in different parts of the Earth than by errors in reading the travel times themselves.

EXERCISES

1.1. A geophone response is quoted as 3.28 V $\text{in}^{-1}\ \text{s}^{-1}$. Give the ground motion in units of m s^{-1} corresponding to an output of 1 mV.

1.2. A simple seismic acquisition system uses 14 bits to record the signal digitally as an integer. Calculate the dynamic range. Convert to deciBels (dB) by using the formula

$$\text{dB} = 20 \log_{10} (\text{dynamic range}). \tag{E1.1}$$

(The factor 10 converts 'Bel' to 'deciBel' and the factor 2 converts from amplitude to energy.)

1.3. The geophone in question 1.1 outputs linearly over the range 1 μV–0.1 V. What is its dynamic range in dB?

1.4. You attach the geophone in question 1.1 to the recorder in question 1.2 and set the recorder's gain to 1 digital count/μV. What is the maximum ground motion recorded?

1.5. A more sophisticated recording system has a 19 bit seismic word, one sign bit (polarity), four gain bits (exponent plus its sign), and a 14 bit mantissa (the digits after the decimal point). Calculate the dynamic range in dB.

1.6. What happens to the sensitivity of the more sophisticated recording system for large signals?

Part I

Processing

2

Mathematical preliminaries: the z- and discrete Fourier transforms

Suppose we have some data digitised as a sequence of N numbers:

$$\{a\} = a_0, a_1, a_2, \ldots, a_{N-1}. \tag{2.1}$$

These numbers must be values of measurements made at regular intervals. The data may be sampled evenly in time, such as a seismogram, or evenly in space, such as an idealised gravity traverse. Whatever the nature of the independent variable, it will be called *time* and the sequence will be called a *time sequence*. The mathematics described in this chapter requires only a sequence of numbers. The main tool for analysing discrete data is the discrete Fourier transform or DFT. The DFT is closely related to Fourier series, with which the reader is probably already familiar. Some basic properties of Fourier series are reviewed in Appendix 1. This chapter begins not with the DFT but with the z-transform, a simpler construct that will allow us to derive some fundamental formulae that apply also to the DFT.

2.1 The z-transform

The *z-transform* is made by simply forming a polynomial in the complex variable z using the elements of the time sequence as coefficients:

$$A(z) = a_0 + a_1 z + a_2 z^2 + \cdots + a_{N-1} z^{N-1}. \tag{2.2}$$

Many physical processes have the effect of multiplying the z-transform of a time sequence by another z-transform. The best known of these is the effect of an instrument: the output of a seismometer, for example, is different from the actual ground motion it is supposed to measure. The relationship is one of multiplying the z-transform of the ground motion by another z-transform that depends on the instrument itself. For reasons that will become apparent later, it is physically impossible to make a device that records the incoming time series precisely.

It is therefore instructive to consider operations on the z-transform rather than on the original time series. Operations on the z-transform all have their counterpart in the time domain. For example, multiplying by z gives

$$zA(z) = a_0 z + a_1 z^2 + a_2 z^3 + \cdots + a_{N-1} z^N. \tag{2.3}$$

This new z-transform corresponds to the time sequence:

$$\{a_+\} = 0, a_0, a_1, a_2, \ldots, a_{N-1}, \tag{2.4}$$

which is the original sequence shifted one space in time, or by an amount equal to the sampling interval Δt. Likewise, multiplying by a power of z, z^k, shifts the time sequence by $k\Delta t$. Here z is called the *unit delay operator*.

In general, multiplication of two z-transforms is equivalent in the time domain to a process called *discrete convolution*. This *discrete convolution theorem* is the most important property of the z-transform. Consider the product of $A(z)$ with $B(z)$, the z-transform of a second time sequence $\{b\}$, whose length M is not necessarily the same as that of the original sequence

$$C(z) = A(z)B(z) = \sum_{k=0}^{N-1} a_k z^k \sum_{l=0}^{M-1} b_l z^l. \tag{2.5}$$

To find the time sequence corresponding to this z-transform we must write it as a polynomial and find the general term. Setting $p = k + l$ to replace subscript l and changing the order of the summation does exactly this:

$$\sum_{k=0}^{N-1} \sum_{l=0}^{M-1} a_k b_l z^{k+l} = \sum_{p=0}^{M+N-2} \sum_{k=0}^{p} a_k b_{p-k} z^p. \tag{2.6}$$

This is just the z-transform of the time sequence $\{c\}$, where

$$c_p = \sum_{k=0}^{p} a_k b_{p-k}. \tag{2.7}$$

$\{c\}$ is the discrete convolution of $\{a\}$ and $\{b\}$; it has length $N + M - 1$, one less than the sum of the lengths of the two contributing sequences.

Henceforth the curly brackets on sequences will be omitted unless this could lead to ambiguity. The convolution is usually written as

$$c = a * b \tag{2.8}$$

and the discrete convolution theorem for z-transforms will be represented by the notation

$$\begin{aligned} a &\leftrightarrow A(z) \\ b &\leftrightarrow B(z) \\ a * b &\leftrightarrow A(z)B(z). \end{aligned} \tag{2.9}$$

As an example, consider the two sequences

$$\begin{aligned} a &= 1, 2, 3, 1, 2 \quad && N = 5 \\ b &= 1, 3, 5, 7 && M = 4. \end{aligned} \tag{2.10}$$

The convolution c will have length $N + M - 1 = 8$ and is given by formula (2.7). The formula must be applied for each of eight values of p:

$$\begin{aligned} c_0 &= a_0 b_0 = 1 \\ c_1 &= a_0 b_1 + a_1 b_0 = 5 \\ c_2 &= a_0 b_2 + a_1 b_1 + a_2 b_0 = 14 \\ c_3 &= a_0 b_3 + a_1 b_2 + a_2 b_1 + a_3 b_0 = 27 \\ c_4 &= a_1 b_3 + a_2 b_2 + a_3 b_1 + a_4 b_0 = 34 \\ c_5 &= a_2 b_3 + a_3 b_2 + a_4 b_1 = 32 \\ c_6 &= a_3 b_3 + a_4 b_2 = 17 \\ c_7 &= a_4 b_3 = 14. \end{aligned} \tag{2.11}$$

The procedure is visualised in Figure 2.1. The elements of the first sequence (a) are written down in order and those of the second sequence (b) are written down in reverse order (because of the minus sign of the k subscript appearing in (2.7)). To compute the first element, c_0, the start of the b sequence is aligned at $k = 0$. The corresponding elements of a and b are multiplied across and summed. Only b_0 lies below an element of the a sequence in the top frame in Figure 2.1, so only the product $a_0 b_0$ contributes to the first term of the convolution. In general, c_p is found by aligning b_0 at $k = p$, multiplying corresponding elements, and summing. It is often the case that one sequence is much longer than the other, the long sequence representing the seismogram and the short sequence an operation to be performed on the seismogram. Think of the short sequence as a 'train' running along the 'rails' of the long sequence as the subscript p is increased.

If multiplication by a z-transform corresponds to convolution, division by a z-transform corresponds to the opposite procedure, *deconvolution*, since $B(z) = C(z)/A(z)$. Inspection of the first equation in example (2.11) shows that we can

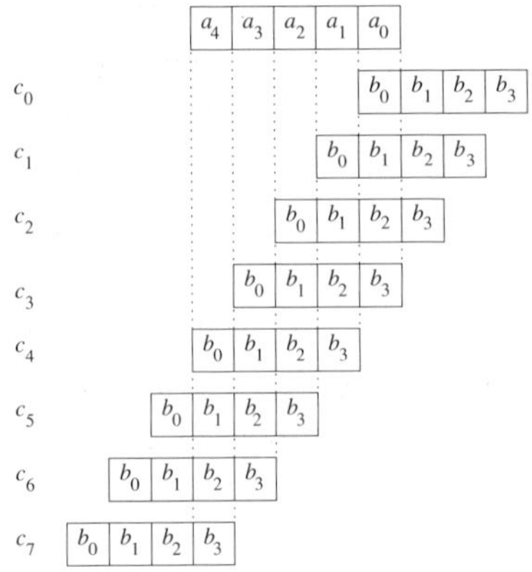

Fig. 2.1. The convolution process. The first sequence a is plotted against subscript k. The second sequence b is reversed and shifted so that b_0 is at $k = p$. Products of the adjacent elements of the two sequences are then summed to give the convolution for that value of p.

find b_0 if $a_0 \neq 0$:

$$b_0 = c_0/a_0. \tag{2.12}$$

Knowing the first term b_0 we can use it to find the next term of the b sequence from the second of (2.11):

$$b_1 = (c_1 - a_1 b_0)/a_0 \tag{2.13}$$

and so on. The general formula to invert (2.7) is obtained by moving the $k = 0$ term in the sum for c_p to the left-hand side and rearranging to give

$$b_p = \frac{\left(c_p - \sum_{k=1}^{p} a_k b_{p-k}\right)}{a_0}. \tag{2.14}$$

This procedure is called *recursion* because the result of each equation is used in all subsequent equations. Notice the division by a_0 at every stage: deconvolution is impossible when $a_0 = 0$. Since $B(z) = C(z)/A(z)$, deconvolution is also impossible when $A(z) = 0$. Deconvolution is discussed in detail in Section 4.3.

We may now define a generalised operation corresponding to multiplication of the z-transform by a rational function (ratio of two polynomials):

$$r(z) = \frac{f(z)}{g(z)} = \frac{(z - z_0)(z - z_1)\ldots(z - z_{N-1})}{(z - p_0)(z - p_1)\ldots(z - p_{M-1})}.$$

In the time domain this corresponds to a combination of convolution and recursion.

The operation describes the relationship between the output of a seismometer and the ground motion, or *instrument response*. Instrument responses of modern seismometers are usually specified in terms of the roots of the two polynomials, the poles and zeroes of the complex z-transform $r(z)$. An ideal seismometer has $r(z) = 1$ and records the ground motion exactly. Some modern seismometers approach this ideal and have rather simple instrument responses; older instruments (and others not designed to record all aspects of the ground motion) have more complicated responses. The SRO (which stands for Seismological Research Observatory) was designed in the 1970s for the global network. Its full response was properly described by no fewer than 14 zeroes and 28 poles. Both input and output are made up of real numbers, so the z-transforms have real coefficients and therefore the roots (poles and zeroes) either are real or occur in complex conjugate pairs.

2.2 The discrete Fourier transform

Setting $z = e^{-i\omega\Delta t}$ into the formula for the z-transform (2.2) and normalising with a factor N gives

$$A(\omega) = \frac{1}{N}\sum_{k=0}^{N-1} a_k e^{-i\omega k\Delta t}. \tag{2.15}$$

This equation is a complex Fourier series (Appendix 1); it defines a continuous function of angular frequency ω that we discretise by choosing

$$\omega_n = \frac{2\pi n}{T} = \frac{2\pi n}{N\Delta t} = 2\pi n\Delta\nu, \tag{2.16}$$

where $\Delta\nu$ is the *sampling frequency*. Substituting into (2.15) and using $T = N\Delta t$ gives the *discrete Fourier transform* (DFT)

$$\begin{aligned} A_n &= A(\omega_n) \\ &= \frac{1}{N}\sum_{k=0}^{N-1} a_k e^{-2\pi i nk/N}; \\ n &= 0, 1, 2, \ldots, N-1. \end{aligned} \tag{2.17}$$

This equation transforms the N values of the time sequence $\{a_k\}$ into another sequence of numbers comprising the N Fourier coefficients $\{A_n\}$. The values of z that yield the DFT are uniformly distributed about the unit circle in the complex plane (Figure 2.2).

An inverse formula exists to recover the original time sequence from the Fourier coefficients:

$$a_k = \sum_{n=0}^{N-1} A_n e^{2\pi i kn/N}. \tag{2.18}$$

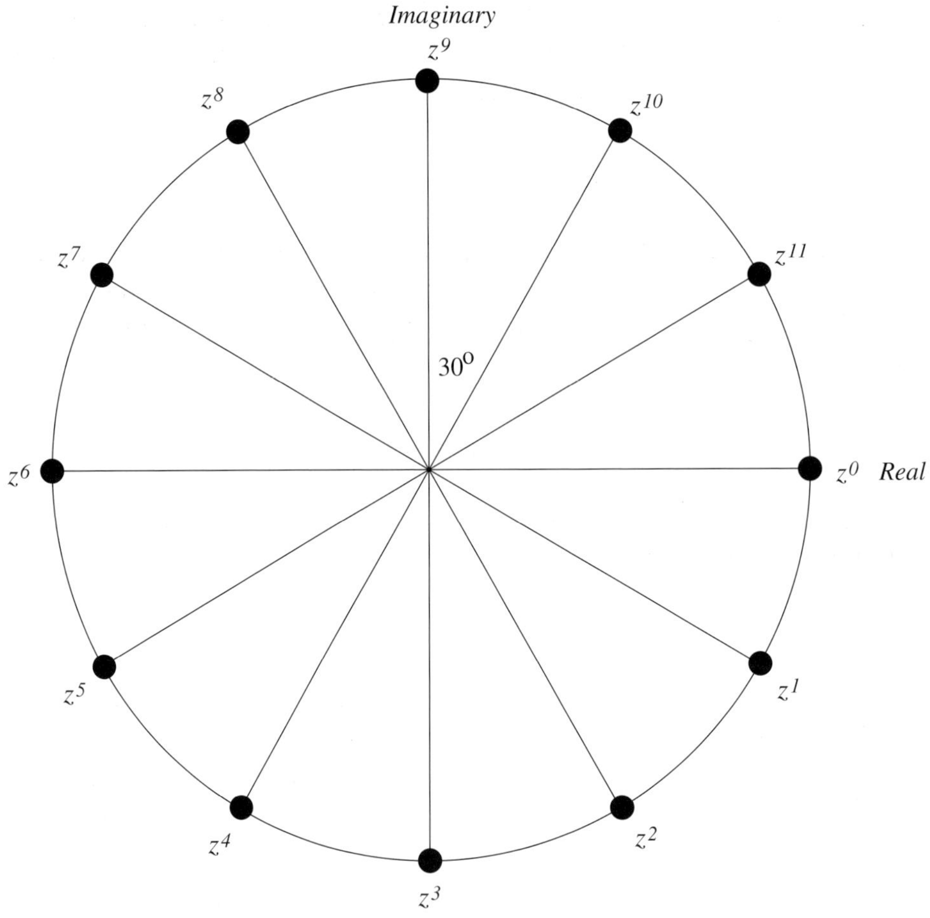

Fig. 2.2. In the Argand diagram our choice for z, $\exp -\mathrm{i}\omega\Delta t$, lies always on the unit circle because $|z| = 1$. Discretisation places the z-points uniformly around the unit circle. In this case $\omega\Delta t = -\pi/12$ and there are 12 points around the unit circle.

To verify this, substitute A_n from (2.17) into the right-hand side of (2.18) to give

$$\sum_{n=0}^{N-1} \frac{1}{N} \sum_{l=0}^{N-1} a_l \mathrm{e}^{-2\pi\mathrm{i}nl/N} \mathrm{e}^{2\pi\mathrm{i}kn/N} = \sum_{l=0}^{N-1} a_l \frac{1}{N} \sum_{n=0}^{N-1} \mathrm{e}^{-2\pi\mathrm{i}n(l-k)/N}. \tag{2.19}$$

The sum over n is a geometric progression. Recall the formula for the sum of a geometric progression, which applies to both real and complex sequences:

$$\sum_{k=0}^{N-1} r^k = \frac{1-r^N}{1-r}. \tag{2.20}$$

In (2.19) the ratio is $r = \mathrm{e}^{-2\pi \mathrm{i}(k-l)/N}$ and r^N is $\mathrm{e}^{-2\pi \mathrm{i}(k-l)}$, which is unity provided $k \neq l$. The sum in (2.19) is therefore zero. When $k = l$ every term is unity, giving a sum of N. *Nyquist's theorem* follows:

$$\frac{1}{N}\sum_{n=0}^{N-1} \mathrm{e}^{-2\pi \mathrm{i}n(k-l)/N} = \delta_{kl}. \tag{2.21}$$

Substituting into (2.19) and using the substitution property of the Kronecker delta (see Box 2.1) leaves a_k, which proves the result. Equations (2.17) and (2.18) are a transform pair that allow us to pass between the time sequence a and its DFT A and back without any loss of information.

Box 2.1. The Kronecker delta

The Kronecker delta, also called the isotropic tensor of rank 2, is simply an array taking the value zero or one:

$$\begin{aligned}\delta_{kl} &= 0 \quad k \neq l \\ &= 1 \quad k = l.\end{aligned}$$

It is useful in time sequence analysis in defining a spike, a sequence containing a single non-zero entry at time corresponding to element $k = l$.

The *substitution property* of the Kronecker delta applies to the subscripts. When multiplied by another time sequence and summed over k the sum yields just one value of the sequence:

$$\sum_k \delta_{kl} a_k = a_0 \cdot 0 + a_1 \cdot 0 + \cdots + a_l \cdot 1 + \cdots = a_l.$$

The subscript k on the a on the left-hand side has been substituted for the subscript l on the right-hand side.

In equations (2.17)–(2.18) subscript k measures time in units of the sampling interval Δt, so $t = k\Delta t$, up to a maximum time $T = N\Delta t$. The n measures frequency in intervals of $\Delta\nu = 1/T$, so $\nu = n\Delta\nu$, up to a maximum of the sampling frequency

$$\nu_{\mathrm{s}} = N\Delta\nu = 1/\Delta t. \tag{2.22}$$

The complex Fourier coefficients A_n describe the contribution of the particular frequency $\omega = 2\pi n\Delta\nu$ to the original time sequence. Writing A_n in terms of modulus and argument gives

$$A_n = R_n \mathrm{e}^{\mathrm{i}\Phi_n}. \tag{2.23}$$

A signal with just one frequency is a sine wave; R_n is the maximum value of the sine wave and Φ_n defines the initial point of the cycle – whether it is truly a sine, or a cosine or some combination. R_n is real and positive and gives a measure of the amount that frequency contributes to the data; a graph of R_n plotted against n is called the *amplitude spectrum* and its square, R_n^2, the *power spectrum*. Φ_n is an angle and describes the phase of this frequency within the time sequence; a graph of Φ_n plotted against n is called the *phase spectrum*.

Consider first the simple example of the DFT of a spike. Spikes are often used in processing because they represent an ideal impulse. The spike sequence of length N has

$$d_k = \delta_{kK}. \tag{2.24}$$

The spike is at time $K\Delta t$. Substituting into (2.17) gives

$$D_n = \frac{1}{N}e^{-2\pi i Kn/N}. \tag{2.25}$$

When $K = 0$, $D_n =$ constant; the spike is at the origin, the phase spectrum is zero and the amplitude spectrum is flat. The spike is said to contain all frequencies equally. Shifting the spike to some other time changes the phase spectrum but not the amplitude spectrum.

An important example of a time sequence is the *boxcar*, a sequence that takes the value 1 or 0 (Figure 2.3). It takes its name from the American term for a railway carriage and its appearance in convolution, when it runs along the 'rails' of the seismogram. The time sequence is defined as

$$\begin{aligned} b_k &= 1 \quad 0 \le k < 10 \\ &= 0 \quad 10 \le k < 100. \end{aligned} \tag{2.26}$$

Substituting into (2.17) gives

$$B_n = \frac{1}{N}\sum_{k=0}^{M-1} e^{-2\pi i kn/N}, \tag{2.27}$$

which is another geometric progression with factor $\exp -2\pi i n/N$ and sum

$$\begin{aligned} B_n &= \frac{1 - e^{-2\pi i n M/N}}{N(1 - e^{-2\pi i n/N})} \\ &= e^{-\pi i n(M-1)/N}\frac{\sin \pi n M/N}{N \sin \pi n/N}. \end{aligned} \tag{2.28}$$

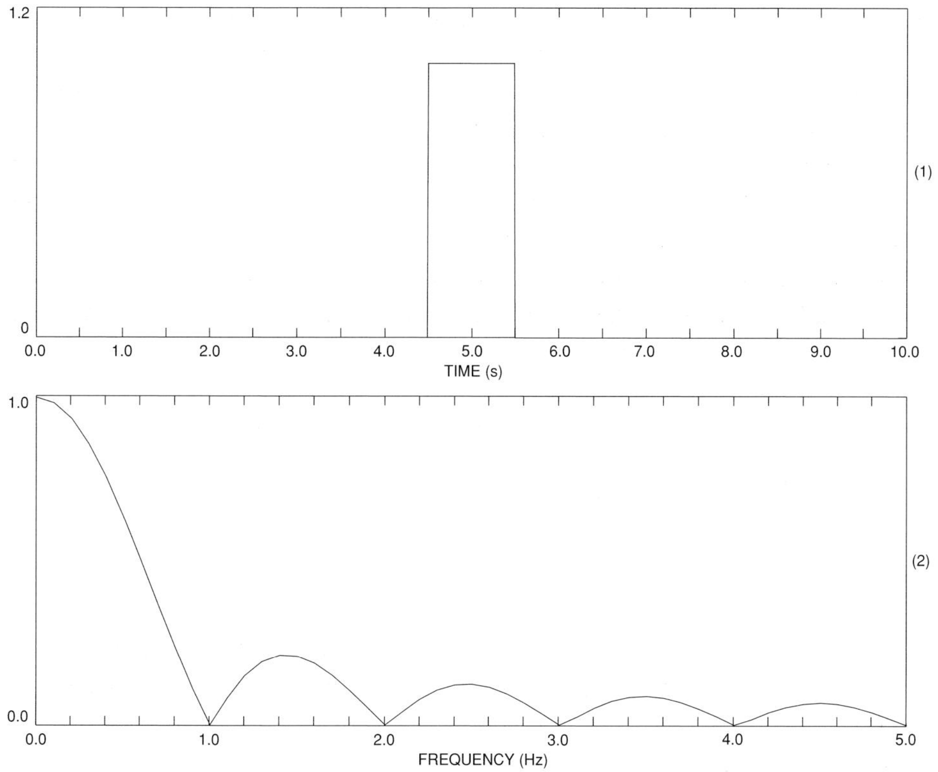

Fig. 2.3. A boxcar sequence for $N = 100$, $M = 10$, and $\Delta t = 0.1$ s. Its amplitude spectrum is shown in the lower trace. The amplitude spectrum remains unchanged because of the shift theorem (Section 2.3.2). Note the zeroes and slow decrease in maximum amplitude from 0 to 5.0 Hz caused by the denominator in equation (2.29).

The amplitude and phase spectra are, from (2.23):

$$R_n = \left| \frac{\sin \pi n M/N}{N \sin \pi n/N} \right| \tag{2.29}$$

$$\Phi_n = -\frac{\pi n \,(M-1)}{N} - \epsilon\pi, \tag{2.30}$$

where ϵ takes the values 0 or 1, depending on the sign of $\sin \pi n M/N$.

The numerator in (2.28) oscillates as n increases. Zeroes occur at $n = N/M$, $2N/M, \ldots$, where M is the length of the (non-zero part of the) boxcar. The denominator increases from zero at $n = 0$ to a maximum at $n = N/2$ then decreases to zero at $n = N$; it modulates the oscillating numerator to give a maximum (height M) at $n = 0$ to a minimum in the centre of the range $n = N/2$. The amplitude spectrum is plotted as the lower trace in Figure 2.3. The larger the value of M, the narrower the central peak of the transform. This is a general property of Fourier transforms (see Appendix 2 for a similar result): the spike has the narrowest time spread and the broadest Fourier transform.

Inspection of (2.28) shows that the phase spectrum is a straight line decreasing with increasing n (frequency). The phase spectrum is ambiguous to an additive integer multiple of 2π; it is sometimes plotted as a continuous straight line and sometimes folded back in the range $(0, 2\pi)$ in a sawtooth pattern.

Cutting out a segment of a time sequence is equivalent to multiplying by a boxcar – all values outside the boxcar are set to zero, while those inside are left unchanged. This is why the boxcar arises so frequently in time series analysis, and why it is important to understand its properties in the frequency domain.

2.3 Properties of the discrete Fourier transform

2.3.1 Relation to continuous Fourier series and Fourier integral transform

There exist analogies between the DFT on the one hand and Fourier series and the Fourier integral transform on the other[1]. Fourier series and the FIT are summarised in Appendices 1 and 2, but this material is not a prerequisite for the present chapter. The forward transform, (2.17), has exactly the same form as the complex Fourier series (A1.10). The coefficients of the complex Fourier series are given by the integral (A1.12):

$$A_n = \frac{1}{T}\int_0^T a(t)\mathrm{e}^{-\mathrm{i}n\omega t}\,\mathrm{d}t. \tag{2.31}$$

Setting $\omega = 2\pi/T$, $T = N\Delta t$, $t = k\Delta t$, where Δt is the sampling interval, and using the trapezium rule to approximate the integral leads to (2.17), the DFT equation for A_n.

Likewise, Fourier integrals may be approximated by the trapezium rule to give sums similar to (2.17) and (2.18). Results for continuous functions from FITs and Fourier series provide intuition for the DFT. For example, the FIT of a Gaussian function $\mathrm{e}^{-\alpha t^2}$ is another Gaussian, $(1/2\sqrt{\pi\alpha})\mathrm{e}^{-\omega^2/4\alpha}$ (see Appendix 2). This useful property of the Gaussian (the only function that transforms into itself) does not apply to the DFT because the Gaussian function never goes to zero; there are end effects, but Gaussian functions are still used in time series analysis. Note also that when α is large the Gaussian is very narrow and sharply peaked, but the transform is very broad. This is another example of a narrow function transforming to

[1] This subsection may be omitted if the reader is unfamiliar with Fourier series or the integral transform.

a broad one, a general property of the Fourier integral transform as well as the DFT.

The Fourier series of a monochromatic sine wave has just one non-zero coefficient, corresponding to the frequency of the sine wave, and the Fourier integral transform of a monochromatic function is a Dirac delta function (Appendix 2) centred on the frequency of the wave. The same is approximately true for the DFT. It will contain a single spike provided the sequence contains an exact number of wavelengths. Differences arise because of the discretisation in time and the finite length of the record. Fourier series are appropriate for analysing periodic functions and FITs are appropriate for functions that continue indefinitely in time. The DFT is appropriate for sequences of finite length that are equally spaced in time, which are our main concern in time sequence analysis.

2.3.2 *Shift theorem*

The DFT is a special case of the z-transform, so multiplication by z $(= e^{-i\omega\Delta t})$ will delay the sequence by one sampling interval Δt. Put the other way, shifting the time sequence one space will multiply the DFT coefficient A_n by $e^{-2\pi i n/N}$. The corresponding amplitude coefficient R_n remains unchanged but the phase is retarded by $2\pi n/N$. This is physically reasonable; delaying in time cannot change the frequency content of the time sequence, only its phase. The spike provides a simple illustration of the shift theorem. Equation (2.25) with $K = 0$ gives $D_n = 1/N$; shifting in time by $K\Delta t$ introduces the phase factor $e^{-2\pi i K/N}$ by the shift theorem, which agrees with (2.25).

2.3.3 *Time reversal*

Reversing the order of a sequence is equivalent to reversing the time and shifting all terms forward by one period (the full length of the sequence). Denote the time-reversed sequence by a' so that $a'_k = a_{N-k}$. The right-hand side of equation (2.17) then gives, with the substitution $l = N - k$,

$$\frac{1}{N}\sum_{k=0}^{N-1} a_{N-k}e^{-2\pi nk/N} = \frac{1}{N}\sum_{l=0}^{N-1} a_l e^{-2\pi n(N-l)/N} = A_n^*. \qquad (2.32)$$

Time reversal therefore complex conjugates the DFT.

Time reversal yields the z-transform of z^{-1} with a delay factor:

$$A'(z) = a_{N-1} + a_{N-2}z + \cdots + a_0 z^{N-1} = z^{N-1}A\left(\frac{1}{z}\right). \qquad (2.33)$$

2.3.4 Periodic repetition

Replace k with $k+N$ in (2.18) to give

$$a_{k+N} = \sum_{n=0}^{N-1} A_n e^{2\pi i(k+N)n/N} = a_k. \qquad (2.34)$$

Similar substitutions show that $a_{k-N} = a_k$, $a_{k+2N} = a_k$, and so forth. The DFT always 'sees' the data as periodically repeated even though we have not specifically required it; the inverse transform (2.18) is only correct if the original data are repeated with period N.

2.3.5 Convolution theorem

The DFT is a special case of the z-transform and we would expect it to obey the convolution theorem provided z has the same meaning in both series and we select the same frequencies ω_n, which means the same sampling interval and number of samples for each series. There is one important difference resulting from the periodic repetition implicit in the DFT. In deriving (2.7) we omitted terms in the sum involving elements of a or b with subscripts which lay outside the specified ranges, 0 to $N-1$ for a and 0 to $M-1$ for b. When the sequences repeat periodically this is no longer correct. We can make the convolution theorem work in its original form by adding zeroes to extend both sequences to length $N+M+1$ or more. This is called 'padding' with zeroes. There is then no overlap with the periodically repeated parts of the sequences (Figure 2.4).

The effect of convolution is often difficult to understand intuitively, whereas the equivalent process of multiplication in the frequency domain can be simpler. Consider the example of convolution with a boxcar. This is a *moving average* (provided the boxcar has height $1/M$). A moving average might be thought suitable for smoothing out high frequencies in the data, but this is not the case. Convolution with b in the time domain is equivalent to multiplication by B in the frequency domain. The amplitude spectrum in Figure 2.3 shows that the procedure will completely eliminate the frequencies $N/M, 2N/M, \ldots$, but will not do a very good job of removing other high frequencies. The zeroed frequencies correspond to oscillations with an exact integer number of wavelengths within the length of the boxcar. Thus a moving average with a length of exactly 12 h will be very effective in reducing a signal with exactly that period, (thermal effects from the Sun, for example), but will not do a good job for neighbouring frequencies (tides, for example, which do not have periods that are exactly multiples of 12 h). In general we do not know the periods in advance, so we cannot design the right moving average, and

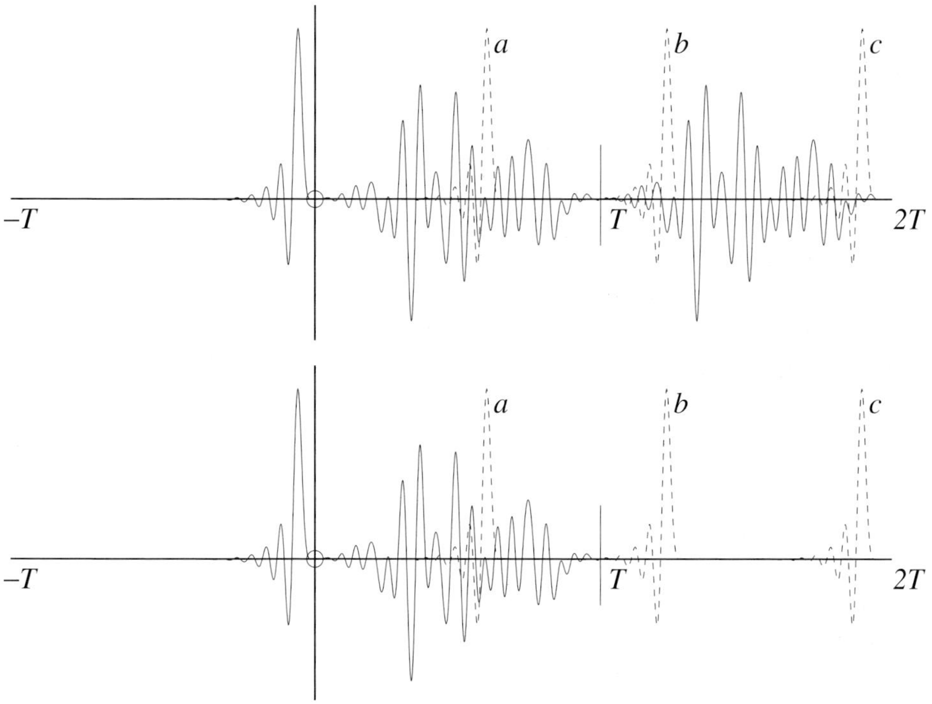

Fig. 2.4. Cyclic convolution of a long trace with a shorter one. The long trace is actually a seismogram and the shorter one is a 'filter' (see Chapter 4.1) designed to enhance specific reflections within the original seismogram. The sequences have been plotted as continuous functions of time, as would normally be done in processing a seismogram. The *a*, *b* and *c* show the location of the shorter filter as it moves through the seismogram. When one time sequence is short, cyclic and ordinary convolutions differ only near the ends. They can be made identical by adding zeroes to both sequences, as shown in the lower figure. In this example, if the seismogram outlasts the duration of significant ground motion there is little difference between cyclic and standard convolution.

few physical phenomena have a single precise frequency. The reduction of other frequencies is determined mainly by the slowly varying denominator. In Chapter 4 we shall explore more effective ways to remove the higher frequencies.

2.3.6 Cyclic convolution

If no padding is applied, the product of two DFTs give the DFT of the *cyclic convolution*. The same formula (2.7) applies but values of the sequence with subscripts outside the range $(0, N-1)$ are the periodically repeated values rather than zeroes. The cyclic convolution is periodic with period N, the same as the original sequences.

The cyclic convolution for the two sequences a and b defined in (2.10) is quite different from the ordinary convolution. First we make them the same length by adding one zero to b, then apply (2.7) while retaining elements of b with subscripts outside the range (0, 4), which are filled in by periodic repetition (e.g. $b_{-1} = b_4 = 0$, $b_{-3} = b_3 = 5$). Equation (2.11) is replaced by

$$\begin{aligned}
c_0 &= a_0b_0 + a_1b_{-1} + a_2b_{-2} + a_3b_{-3} + a_4b_{-4} = 33 \\
c_1 &= a_0b_1 + a_1b_0 + a_2b_{-1} + a_3b_{-2} + a_4b_{-3} = 22 \\
c_2 &= a_0b_2 + a_1b_1 + a_2b_0 + a_3b_{-1} + a_4b_{-2} = 28 \\
c_3 &= a_0b_3 + a_1b_2 + a_2b_1 + a_3b_0 + a_4b_{-1} = 27 \\
c_4 &= a_0b_4 + a_1b_3 + a_2b_2 + a_3b_1 + a_4b_0 = 34.
\end{aligned} \tag{2.35}$$

The convolution is of length 5, the same as both contributing series, and repeats periodically with period 5. It is quite different from the non-cyclic convolution in (2.11): (1, 5, 14, 27, 34, 32, 17, 14).

2.3.7 Differentiation and integration

Differentiation and integration only really apply to continuous functions of time, so first let us recover a continuous function of time from our time sequence by setting $t = k\Delta t$ and $\omega_n = 2\pi n/N\Delta t$ in (2.18):

$$a(t) = \sum_{n=0}^{N-1} A_n e^{i\omega_n t}. \tag{2.36}$$

Differentiating with respect to time gives

$$\frac{da}{dt} = \sum_{n=0}^{N-1} i\omega_n A_n e^{i\omega_n t}. \tag{2.37}$$

Now discretise this equation by setting $t = k\Delta t$ and

$$\dot{a}_k = \left.\frac{da}{dt}\right|_{t=k\Delta t}$$

to give

$$\dot{a}_k = \sum_{n=0}^{N-1} i\omega_n A_n e^{2\pi ikn/N}. \tag{2.38}$$

This is exactly the same form as the inverse DFT (2.18), so $\dot{a}_k$ and $i\omega_n A_n$ must be transforms of each other.

Differentiation with respect to time is therefore equivalent to multiplication by frequency (and a phase factor i) in the frequency domain; integration with respect to time is equivalent to division in the frequency domain (plus an arbitrary constant of integration). The same properties hold for the FIT (Appendix 2).

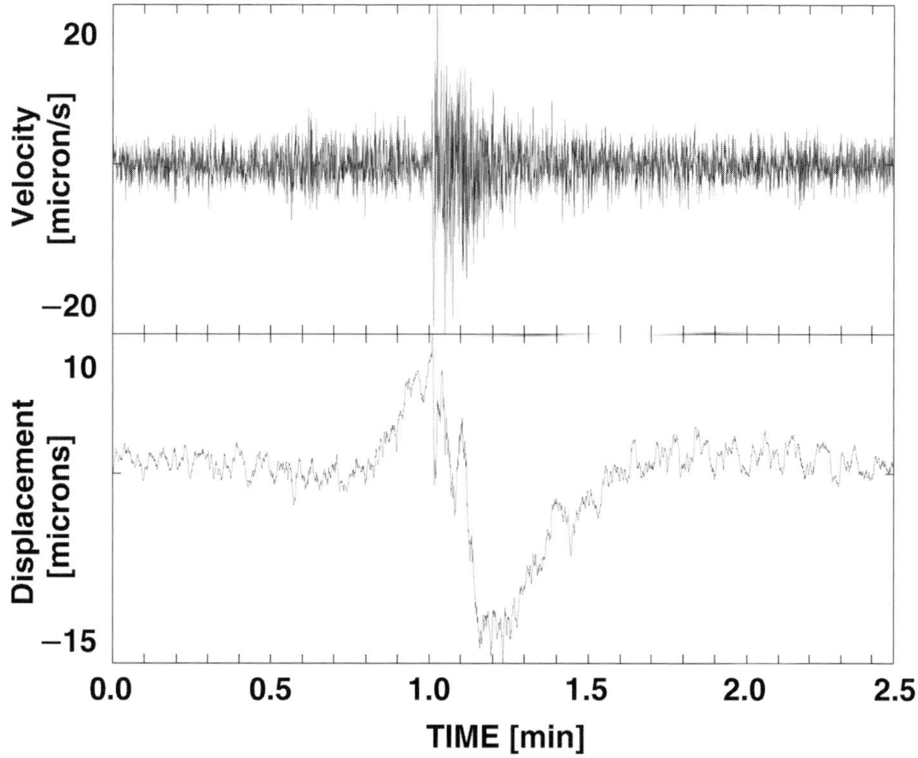

Fig. 2.5. Broadband velocity (upper trace) and displacement (lower trace) seismograms obtained from the same record on Stromboli volcano. A depression is most noticeable in the displacement; the displacement trace also has much less high-frequency noise because of the integration. Reproduced, with permission, from Neuberg and Luckett (1996).

For example, we sometimes need to differentiate and integrate seismograms to convert between displacement, velocity, and acceleration. The change in appearance of the seismograms can sometimes be quite dramatic because multiplication by frequency enhances higher frequencies relative to lower frequencies; thus the velocity seismogram can appear noisier than the displacement seismogram (Figure 2.5).

2.3.8 Parseval's theorem

Substituting for A_n from (2.17) gives:

$$|A_n|^2 = \frac{1}{N^2} \sum_{k=0}^{N-1} a_k e^{2\pi i nk/N} \sum_{l=0}^{N-1} a_l e^{-2\pi i nl/N}. \tag{2.39}$$

Summing over n using the Nyquist theorem and substitution property of the Kronecker delta gives:

$$\begin{aligned}\sum_{n=0}^{N-1}|A_n|^2 &= \frac{1}{N^2}\sum_{k,l}a_k a_l \sum_{n=0}^{N-1}e^{-2\pi i n(l-k)/N} \\ &= \frac{1}{N}\sum_{k,l}a_k a_l \delta_{kl} \\ \sum_{n=0}^{N-1}|A_n|^2 &= \frac{1}{N}\sum_{k=0}^{N-1}|a_k|^2 .\end{aligned} \tag{2.40}$$

This is called Parseval's theorem; it ensures equality of 'energy' between the time and frequency domains.

2.3.9 Fast Fourier transform

Equation (2.17) for the DFT requires a sum over N terms for each of N frequency values n; the total computation required is therefore N^2 complex multiplications and additions (operations), assuming all the exponential factors have been computed in advance. A clever algorithm, called the *fast Fourier transform* or FFT, accomplishes the same task with substantially less computation. To illustrate the principle of the FFT, suppose N is divisible by 2 and split the sum (2.17) into two parts:

$$\begin{aligned}NA_n &= \sum_{k=0}^{N-1}a_k e^{-2\pi i n k/N} \\ &= \sum_{k=0}^{N/2-1}a_k e^{-2\pi i n k/N} + \sum_{k=N/2}^{N-1}a_k e^{-2\pi i n k/N} \\ &\sum_{k=0}^{N/2-1}(a_k + a_{k+N/2}e^{-\pi i n})e^{-2\pi i n k/N} .\end{aligned} \tag{2.41}$$

This sum is achieved by first forming the combinations $a_k + a_{k+N/2}e^{-\pi i n}$ ($N/2$ operations) and then performing the sum of $N/2$ terms. The total number of operations for all frequency terms is now $N \cdot N/2 + N/2$, a reduction of almost half for large N when the additional $N/2$ operations may be neglected. There is no need to stop there: provided N can be divided by 2 again we can divide the sum in half once more to obtain a further speed up of almost a factor of two. When N is a power of 2 the sum may be divided $\log_2 N$ times, with a final operation count for large N of about $4N \log_2 N$.

The modern version of the FFT algorithm dates from Cooley and Tukey (1965). It was known much earlier but did not find widespread application until the advent of digital computers led to large-scale data processing. The power of 2 fast algorithm is the most widely used, but fast algorithms are available for N containing larger prime factors.

The speed advantage of the FFT is enormous. Consider the case $N = 10^6 \approx 2^{20}$ and a slow computer performing a million operations per second. The conventional sum requires $N^2 = 10^{12}$ operations, which take 10^6 s or 11.5 days to do the transform. The FFT requires $4 \times 20 \times 10^6 = 8 \times 10^7$ operations or about 80 s! An 80 s computation could be used repeatedly in a data processing experiment, while an 11.5 day computation is only practical for a one-off calculation.

The FFT not only made it practical to compute spectra for large datasets, it also reduced the time required to convolve two sequences of roughly equal length using the convolution theorem. We transform both sequences with the FFT, multiply the transforms together, then transform back using the FFT. Despite its apparent clumsiness, this procedure is faster than using equation (2.7). Like a conventional ('slow') DFT, convolution using equation (2.7) also requires $O(N^2)$ operations when $N = M$ and is large. Using the convolution theorem, we pad each sequence with zeroes to avoid the problems of periodic repetition, take the DFT of both ($16N \log_2 N$ operations), multiply the transforms together (N operations) and inverse transform ($8N \log_2 N$ operations), to produce the same result. If $N > 24 \log_2 N$ the transform method will be faster. When one sequence is much longer than the other, $N \gg M$, the time domain calculation will usually be faster.

2.3.10 Symmetry of the transform

Equations (2.17) and (2.18) give a means of transforming a sequence of length N into another sequence of the same length and back again without any loss of information. The two formulae are quite similar, the only differences being the normalising factor $1/N$ in (2.17) and the sign of the exponent, both of which have been chosen arbitrarily by convention. The mathematics works equally well for exponents with the other sign convention, provided the signs are different in the two equations and we apply them consistently, or for a normalisation factor N in front of (2.18) instead of $1/N$ in front of (2.17). Some authors use a normalising factor $1/\sqrt{N}$, which gives a symmetric transform pair with the same factor in each formula. Whatever the convention, it is fixed by the definition of A_n.

This symmetry means that, apart from signs and normalisations, any statement we make about the time sequence and its transform can also be made about the transform and its time sequence. For example, we found the DFT of a boxcar to be a ratio of sines. Symmetry requires that the DFT of a time sequence made up of the ratio of sines would be a boxcar. Another example is given by the convolution theorem:

$$a * b \leftrightarrow AB,$$

which states that convolution in the time domain is multiplication in the frequency domain. It follows that convolution in the frequency domain is multiplication in the time domain:

$$A * B \leftrightarrow ab.$$

The properties of the discrete Fourier transform are summarised is Box 2.2.

Box 2.2. Summary: properties of the discrete Fourier transform

(i) Forward and inverse transforms:

$$A_n = \frac{1}{N}\sum_{k=0}^{N-1} a_k e^{-2\pi i k n/N}$$

$$a_k = \sum_{n=0}^{N-1} A_n e^{2\pi i k n/N}.$$

(ii) Amplitude and phase spectra: $R_n = |A_n|$; $\Phi_n = \arg A_n$; $A_n = R_n \exp i\Phi_n$.

(iii) A_0 = average of the data, the zero-frequency component, or 'DC shift'.

(iv) Maximum frequency $1/\Delta t$.

(v) Periodic repetition: $A_{N+n} = A_n$.

(vi) Shift theorem: $a_{k+M} \leftrightarrow A_n e^{-2\pi i n M/N}$.

(vii) $N\delta_{i0} \leftrightarrow 1$ (all n);
$N\delta_{iM} \leftrightarrow A_n e^{-2\pi i n M/N}$.

(viii) Time reversal $\leftrightarrow$ complex conjugation.

(ix) Differentiation and integration:
$\dot{a}(t) \leftrightarrow i\omega_n A_n$;
$\int a(t)\mathrm{d}t \leftrightarrow A_n/(i\omega_n) + \text{constant}$.

(x) Convolution theorem: $a * b \leftrightarrow AB$.

(xi) Convolution with a spike returns the original sequence shifted.

2.4 DFT of random sequences

Noise in a time sequence should sometimes be represented as a random variable, representing a quantity that is itself unpredictable but which has known or knowable statistical properties. The expectation operator $\mathcal{E}$ is applied to functions of a random variable to yield its statistical properties. The expectation is normally taken to be the average of many realisations of the variable. A random variable X has an *expected value* $\mathcal{E}(X)$ and *variance* $\mathcal{V}(X) = \mathcal{E}[(X - \mathcal{E}(X))^2]$ derivable from its probability distribution. For two random variables X and Y we can define a *covariance*

$$\mathcal{C}(X, Y) = \mathcal{E}[(X - \mathcal{E}(X))(Y - \mathcal{E}(Y))]. \tag{2.42}$$

Note that $\mathcal{C}(X, X) = \mathcal{V}(X)$. The covariance measures the dependence of X and Y on each other; if they are independent then $\mathcal{C}(X, Y) = 0$ and they are said to be *uncorrelated*. Given a collection of N random variables $(X_0, X_1, X_2, \ldots, X_{N-1})$ we may define a *covariance matrix*:

$$\mathsf{C}_{ij} = \mathcal{C}\left(X_i, X_j\right). \tag{2.43}$$

For independent random variables the covariance matrix is diagonal with variances on the diagonal. Covariances are discussed in more detail in Chapter 4.

In time series analysis the expectation operator may be replaced by an average over time provided the statistical properties do not vary with time. Such a sequence is called *stationary*. We define a stationary stochastic sequence $(X_0, X_1, X_2, \ldots, X_{N-1})$ as one whose covariances are dependent on the time difference but not the time itself, i.e. $\mathcal{C}(X_k, X_{k+p})$ depends on suffix p but not suffix k. Suppose the sequence $\{a_k;\ \ k = 0, 1, \ldots, N - 1\}$ is one realisation of this stochastic sequence and the X_k have zero mean. Replacing the expectation operator in the covariance with a time average gives an estimate of the covariances

$$\phi_p = \frac{1}{2N - 1} \sum_k a_k a_{k+p}. \tag{2.44}$$

The sequence ϕ_p is called the *autocorrelation* and will be discussed further in Section 4.2.

The DFT of the autocorrelation is the power spectrum:

$$\phi_p \leftrightarrow |A_n|^2. \tag{2.45}$$

The result follows from the convolution theorem and is not that hard to prove. It is given more completely in Section 4.2. The statistical properties of the power spectrum of a stationary stochastic process can therefore be estimated because the power spectrum can be derived directly from the covariances. This allows us to manipulate the spectra of random signals and puts time series analysis on a firm statistical footing. A practical example of noise estimation is given in Section 3.3.3.

A useful random noise sequence n has uncorrelated individual elements. The covariances are therefore zero for all lags p except for $p = 0$, and the autocorrelation approximates to a spike at the origin. The DFT of a spike at the origin is a constant, so the expected values of the power spectrum of n are all the same constant. Such a spectrum is said to be *white* because it contains all frequencies equally, and the random noise sequence n is said to be a sample of *white noise*.

EXERCISES

2.1. Write down the z-transform $A(z)$ of the sequence

$$a_k = \{6, -5, 1).$$

Obtain the first three terms of the time sequence corresponding to the inverse of this z-transform $A^{-1}(z)$. (Hint: factorise $A(z)$, use partial fractions on $A^{-1}(z)$, expand each term by the binomial theorem and collect powers of z.) For what values of z does your expression for $A^{-1}(z)$ converge?

2.2. Obtain the convolution of the two time sequences

$$a_k = \{6, -5, 1, 0, -1, 3, -2)$$

and

$$b_k = \left\{\frac{1}{2}, \frac{1}{2}\right\}.$$

Why is this operation called a *moving average*?

2.3. Use recursion to show that the result of deconvolving the time sequence

$$c = (1, 5, 14, 27, 34, 32, 17, 14)$$

with

$$a = (1, 2, 3, 1, 2)$$

is

$$b = (1, 3, 5, 7).$$

Investigate elements of b beyond the given sequence.

2.4. Show that the discrete Fourier transform of the sequence

$$(0, 1, -1, 0)$$

is

$$(A_0, A_1, A_2, A_3) = \left[0, \frac{1}{4}(1-i), -\frac{1}{2}, \frac{1}{4}(1+i)\right].$$

(a) Verify the general relation $A_{N-n} = A_n^*$.
(b) Plot the amplitude and phase spectra.

2.5. Sketch the amplitude and phase spectra of a boxcar window with sampling rate Δt, total length of series N, taper fraction f (i.e. the number of non-zero terms in the boxcar sequence is $M = fN$. Show on your plot the values of n at the zeroes of the amplitude spectrum. You may assume $1/f$ is an integer.

Obtain the ratio of the values of the amplitude spectrum at $nf = 1.5$ to the central peak at $n = 0$. Find the numerical value of this ratio for $M = 10$.

2.6. Find the cyclic convolution of the two time sequences

$$a_k = (6, -5, 1, 0, -1, 3, -2)$$

and

$$b_k = \frac{1}{2}(1, 1, 0, 0, 0, 0, 0).$$

Compare your result with the ordinary convolution obtained in question 2.2.

2.7. Find the DFT of the time sequence $a_k = (1, -1, 0, 0)$ by (a) direct calculation and (b) using the shift theorem and your solution to question 2.4.

2.8. Verify Parseval's theorem for the time sequences in questions 2.4 and 2.6 above.

2.9. Find the DFT of the sequence $(1, -1)$. Compare its amplitude spectrum with that of the sequence $(1, -1, 0, 0)$, found in question 2.4, and comment on the effect on the amplitude spectrum of adding zeroes to a time sequence.

2.10. Given that the Fourier series for an inverted parabola on the interval $(0, 1)$ is

$$t - t^2 = \frac{1}{6} - \sum_{n=1}^{\infty} \frac{\cos 2\pi n t}{n^2 \pi^2}$$

find, by differentiation or otherwise, the Fourier series of the sawtooth function t in the same interval.

In both cases sketch the function and its Fourier series on the interval $(-1, 2)$. (See Appendix 1.)

2.11. **Computer exercise: introduction to the `pitsa` program**

`pitsa` performs basic operations on time series. It may be downloaded from the web site given in Appendix 7. The program is designed for seismic traces but can be applied to any time sequence. It is menu-driven, making it particularly suitable for beginners. Here we use some `pitsa` menus to illustrate the theoretical results of this chapter in a practical way.

Generate a set of sinusoidal time series by using

`Utilities` → `Testsignal` → `sin/cos`

with a sampling rate of 20 Hz, length of 51.2 s, and any amplitude. Start with a frequency of 1 Hz and calculate the spectra with

`Advanced tools` → `Spectrum` → `FFT` → `boxcar taper (taper fraction = 0.)`

(Tapering and the boxcar are discussed in the next chapter; at this stage all you need to know is that a zero fraction taper does not alter the signal in any way.) Select the `amplitude spectrum`

(a) Explain why the spectrum looks the way it does.

(b) Now gradually increase the frequency of the sine wave by 1 Hz at a time, up to 9 Hz, taking the spectrum each time. Explain why the spectrum changes in the way it does.

(c) If you want to save the results, use
`Files/Traces` → `Save Files` → `ISAM`
This writes out traces to a compact set of files with the same root name.
If you want hardcopy of the plot, use
`Setup` → `Hardcopy Mode On` the click on `To printer`
if you are set up to print, or better `To numbered file` for a postscript file.

(d) *Time shifts*. Generate a Brune source (a theoretical form of the time function of an earthquake source) with
`Utilities` → `Testsignal` → `Brune source`
Make it `Noise free` and 1024 points. Otherwise use the defaults. Take the spectrum as above, and repeat for the `Phase spectrum`, `Real part`, and `Imaginary part`. You should now have five plots on the screen. Examine the plots of the spectra to see if you can tell if the frequency content and phase are reasonable for the time function, and that the real and imaginary parts constitute the same sequence of complex numbers as the amplitude and phase.
Now time shift the signal with
`Utilities` → `Trace utilities` → `single channel utilities` → `cyclic shift`
and choose a 2 s shift (you need to have remembered the sampling frequency to compute the number of points). Take four components of the spectrum as before. You now have 10 plots on the screen. Compare the spectra of the shifted time series with the original. Explain the differences.

(e) *Time reversal*. Now time reverse the shifted Brune source by
`Utilities` → `Trace utilities` → `Single channel utilities` → `Reverse trace in time`
and shift it back with a `cyclic shift` of 8 s. Now take the four components of the spectrum again. You now have 15 plots on the screen. Compare the spectra and verify that time-shifting complex conjugates the transform.

(f) *Convolution with a spike*. Generate a Ricker wavelet with
`Utilities` → `Testsignal` → `Ricker wavelet`
Use the default parameters and dominant frequency 1 Hz. Generate a noise-free, unit amplitude spike at 2 s with
`Utilities` → `Testsignal` → `spikes`
Now convolve the two in the time domain with `Advanced tools` → `(De)convolution` → `Convolution (TIME)`
Explain the result in terms of theory in the text.

(g) *Convolution in time and frequency*. Now generate a mixed signal (exponentially decaying sine wave) with
`Utilities` → `Testsignal` → `Mixed signal`
with amplitude 1, frequency 1, phase 0, $B = 0.1$ and remaining parameters zero. Convolve the Ricker wavelet with the mixed signal in both time and frequency. The option

`Advanced tools` → `(De)convolution` → `Convolution (FREQ)`
asks for the number of points in the frequency domain. Use both 1024 and 2048. Now note that some of the convolutions go to 10 s and some to 20 s. Get them onto the same scale with
`Utilities` → `Header access` → `Plot` → `Viewport`
and double the second parameter (the maximum value of the x-axis). Compare the three versions of the convolution and explain any differences.

(h) *Noise spectra.* Generate a noisy signal with
`Utilities` → `Trace utilities` → `Constant`
Choose the Gaussian distributed noise with unit amplitude, and zero amplitude for the constant. Take the amplitude spectrum and examine it. Is it what you expect? Repeat with 10 traces (not the same ones, as they will have the same noise values), take their spectra, and sum with
`Utilities` → `Trace utilities` → `Stack Traces` → `Plain sum`
Do you think averaging more and more traces will produce a completely flat spectrum?

3

Practical estimation of spectra

3.1 Aliasing

The transform pair (2.17) and (2.18) are exact and therefore no information is lost in performing the DFT: we can move freely between time and frequency domains as circumstances require. However, before digitising we had a continuous function of time $a(t)$ with a continuous Fourier transform $A(\omega)$. We should therefore explore exactly what information is lost by digitising.

It follows from the periodicity of the DFT that

$$A_{N+n} = A_n. \tag{3.1}$$

For real data we can take the complex conjugate of (2.17) and show also that

$$A_{N-n} = A_n^*. \tag{3.2}$$

Physically, this means that Fourier coefficients for frequencies higher than $N/2$ are determined exactly from the first $N/2 + 1$; above N they repeat periodically and between $N/2$ and N they reflect with the same amplitude and a phase change of π. Equation (2.17) allows us to transform N real values in the time domain into any number of complex values A_n, but these only contain N independent real and imaginary parts. There is an irritating complication with A_0, which is real because it corresponds to zero frequency, and $A_{N/2}$, which is also real because of equation (3.2). If N is even there are therefore two real coefficients and $N/2 - 1$ complex coefficients, giving a total of N independent real numbers in the transform.

Equations (3.1) and (3.2) define *aliasing*, the mapping of higher frequencies onto the range $[0, N/2]$: after digitisation higher frequencies will masquerade as lower frequencies. The highest meaningful coefficient in the DFT is $A_{N/2}$, corresponding to the *Nyquist frequency*:

$$\nu_N = 1/2\Delta t. \tag{3.3}$$

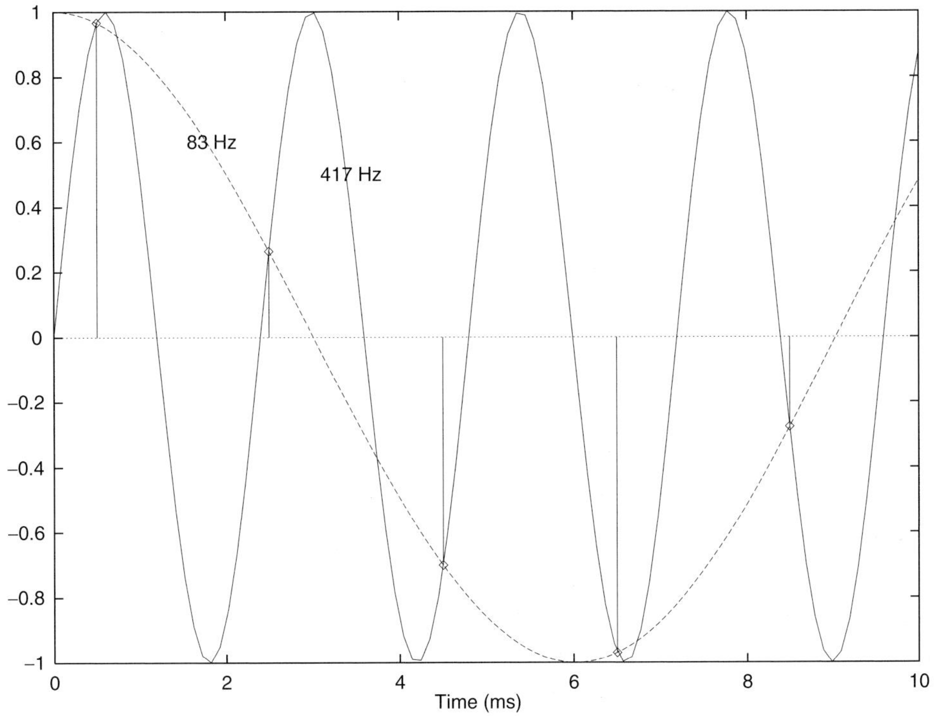

Fig. 3.1. If the sampling interval is too large cycles are skipped in the original signal, giving the impression of a much longer period. Large dots denote values sampled at 500 Hz; they lie on the sine wave with frequency 417 Hz (solid line, the aliased frequency) as well as the sine wave with true frequency 83 Hz (dashed line).

The sine wave associated with $A_{N/2}$ passes through zero at all the digitisation points (Figure 3.1): any component of this sine wave in the original continuous function will not appear in the digitised sequence. Higher frequencies are aliased because cycles are 'missed' by the digitised sequence, as illustrated in Figure 3.1.

A useful illustration of aliasing is the appearance of spokes on a wheel, each an angle α apart, captured on film. A film made from images every, say, 0.025 s (Δt) appears to us to rotate with angular velocity Ω because the spoke moves through an angle $\theta = \Omega \Delta t$ between successive frames. As the wheel speeds up it reaches a point where θ exceeds half the angle between the spokes, and if the spokes are identical we cannot tell from the digitised film whether the wheel is rotating forwards at a rate $\theta / \Delta t$ or backwards at a rate $(\alpha - \theta)/\Delta t$. The brain interprets the image as backwards rotation. Reversal of the sense of rotation corresponds to complex conjugation in (3.2); periodic repetition starts for the faster rotation speed $\Omega = \alpha / \Delta t$, when each spoke comes into line with the next between frames.

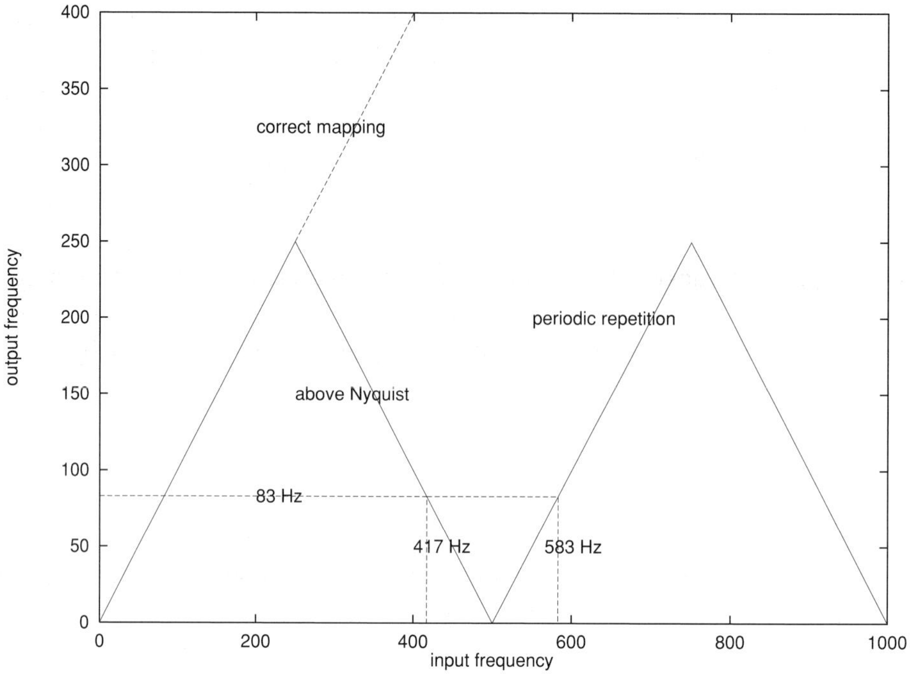

Fig. 3.2. Digitising frequencies higher than the Nyquist frequency $\nu_N = 1/2\Delta t$ back into the range $(0, \nu_N)$. The frequency in the digitised signal is calculated from this graph, with the example in which both frequencies $583 = 500 + 83$ and $417 = 500 - 83$ are both aliased to 83 Hz. The Nyquist frequency is 250 Hz and the sampling interval 2 ms.

Figure 3.2 gives a graphical method for finding the aliased frequency by plotting the frequency of the original continuous function, the input frequency to the digitisation process, against the frequency of the digitised sequence, the aliased frequency. This applies only to the amplitude of the coefficients; there is a phase change of π for each of the negative slope sections in the figure. In this example the sampling interval, Δt, is 2 ms, typical of a seismic reflection survey. The sampling frequency is therefore $\nu_s = 1/\Delta t = 500$ Hz. The Nyquist frequency is half the sampling frequency, or 250 Hz. All frequencies are aliased into the band 0–250 Hz. Equation (3.2) shows that 417 Hz is aliased to $500 - 417 = 83$ Hz, and (3.1) shows that 583 Hz is aliased to $517 - 500 = 17$ Hz.

It should be clear from this example that aliasing can have disastrous effects. Very high frequencies can enter at very low frequency and completely mask the signal of interest. There is no way to undo the effects of aliasing after digitisation: the information allowing us to discriminate frequencies above the ν_N from those below ν_N is lost. It is therefore essential to remove all energy above the Nyquist frequency before digitising. This is usually done by an *anti-alias filter* applied to the analogue signal.

It is sometimes desirable to reduce the volume of data by resampling at a longer interval. For example, a broadband record may have a sampling interval of 0.01 s but a study of surface waves needs a sampling interval of only 1 s. In this case the original digitised sequence must be preprocessed to remove all frequencies above the new Nyquist (0.5 Hz) before making the new sampling.

We have seen that information is lost by digitising at too large a time interval. We now ask a similar question, can we recover the original signal from the digitised samples? The answer is yes, provided the original signal contained no energy above the Nyquist frequency. Reconstruction of the original time series $a(t)$ from its samples $a_k = a(k\Delta t)$ is an interpolation, the precise form being given by *Shannon's sampling theorem*:

$$a(t) = \sum_{k=0}^{K-1} a_k \frac{\sin \pi \nu_N (t - k\Delta t)}{\pi \nu_N (t - k\Delta t)}. \tag{3.4}$$

A proof of Shannon's sampling theorem is given in Appendix 3. Equation (3.4) has the form of a discrete convolution of the digitised time sequence with the sinc function $\sin x/x$, where $x = \pi \nu_N t$, which vanishes at the digitising times $k\Delta t$ except for $t = 0$. The convolution (3.4) performs the interpolation between points in exactly the right way to force frequencies above ν_N to be zero.

Symmetry of the Fourier transform allows us to make the same statement in the frequency domain: if a function of time is truncated ($f(t) = 0$ for $|t| > T$) then its Fourier transform can be recovered from its values at a set of discrete frequencies $n/T = n\Delta\nu$. We sometimes need to add zeroes to a time sequence, for example to increase its length to a power of 2 in order to use an FFT or to improve resolution, and we should know how this changes the spectrum. Consider one time sequence, a, of length M and a second time sequence, c, of length $N > M$, with

$$\begin{aligned} c_k &= a_k; \qquad k = 0, 1, \ldots, M-1 \\ &= 0; \qquad k = M, M+1, \ldots, N-1. \end{aligned} \tag{3.5}$$

Clearly

$$c_k = b_k a_k, \tag{3.6}$$

where b is the boxcar sequence given in equation (2.26). Multiplication in the time domain is convolution in the frequency domain, in this case with the DFT of the boxcar:

$$\begin{aligned} C_n &= \sum_{p=0}^{N-1} B_p A_{n-p} \\ &= \frac{1}{N} \sum_{p=0}^{N-1} \frac{\sin \pi p M/N}{\sin \pi p/N} e^{-\pi i p(M-1)/N} A_{n-p}. \end{aligned} \tag{3.7}$$

This is an interpolation of the A spectrum from M to N points over the same frequency range. It is analogous to the continuous case in (3.4). By adding zeroes to the end of a sequence we have implicitly kept Δt the same, the total length is increased to $N\Delta t$, and the frequency spacing of the transform has decreased to $1/N\Delta t$. 'Padding' by adding zeroes is often a useful thing to do. No information is lost or added, and the additional frequency points can make the plot of the spectrum clearer.

3.2 Spectral leakage and tapering

Examining the spectrum to determine the frequency characteristics of the signal is a useful preliminary to processing. An accurate spectrum is essential if we wish to identify periodicities in the data, the Earth's normal modes for example.

Ideally our data would be a continuous function of time extending forever; the spectrum would then be the Fourier integral transform. In practice we have a discretised series of finite length, which introduces two fundamental limitations on the derived spectrum. The finite length T limits the frequency spacing to $\Delta\nu = 1/T$, and the sampling interval Δt limits the maximum meaningful frequency to ν_N.

In the previous section we found that adding zeroes to a time series reduces the frequency spacing of the spectrum without changing its content. Adding zeroes therefore improves the spectrum's appearance without introducing any spurious new information; it is also a convenient device for lengthening the time series to the next power of two when using the FFT.

The DFT uses a finite number of terms to represent the time series and is therefore susceptible to Gibbs's phenomenon should the original time series contain any discontinuities, just as Fourier series suffer from Gibbs's phenomenon when representing discontinuous functions (Appendix 1). The DFT treats the time series as periodically continued, so a 'hidden' discontinuity will exist if the last value is markedly different from the first: this must be guarded against, usually by reducing the early and late values of the time sequence.

It is useful to consider a finite time sequence lasting time T as a section of an ideal sequence of infinite length. Selecting an interval from a longer time sequence is called *windowing*, and it can be achieved mathematically by multiplying the ideal sequence by a boxcar function that is zero everywhere outside the range $(0, T)$ and equal to unity inside the range.

Suppose we have an ideal time sequence whose spectrum contains some lines and other structure, and window it in time. Multiplication in the time domain is equivalent to convolution in the frequency domain, so we have convolved the ideal spectrum with the DFT of a boxcar, B_n, given in equation (2.28) and shown in Figure 2.2. This function has a central peak and many side lobes. A line in the

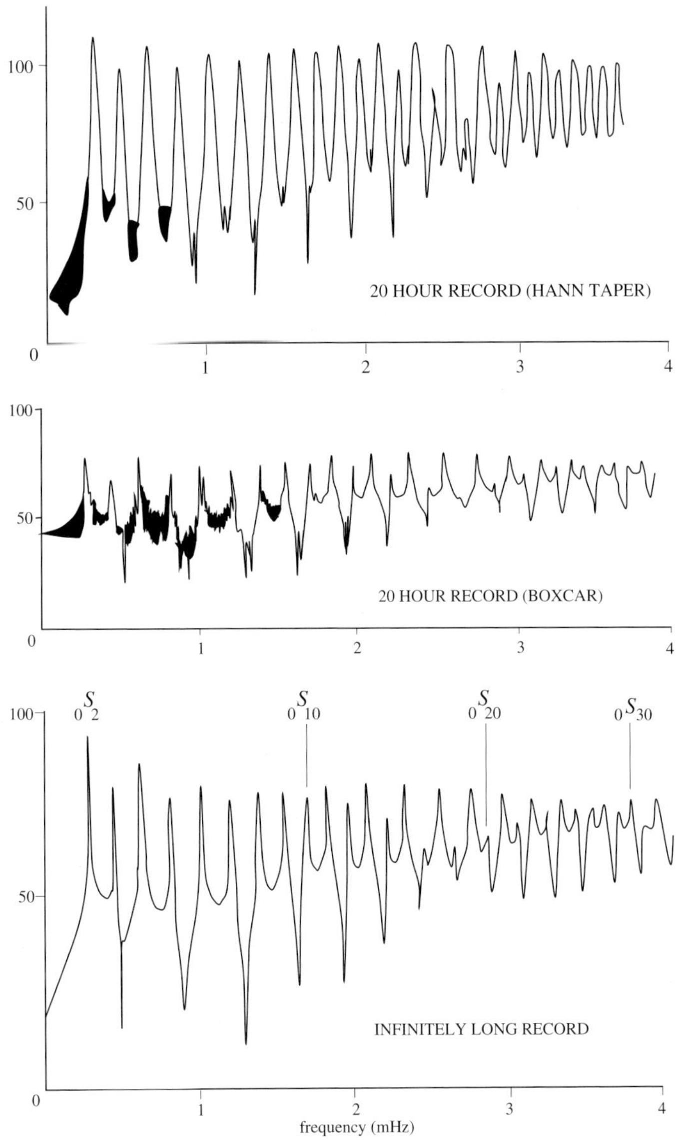

Fig. 3.3. Synthetic boxcar window power spectra for an explosive source recorded at a distance of 95°. Reproduced, with permission, from Dahlen (1982).

ideal spectrum at frequency ω will convolve with B_n to return the same sequence shifted to ω ('convolution with a spike'). A single peak is therefore spread across a range of frequencies with many spurious secondary peaks. This is called *spectral leakage* because windowing in time spreads the energy in the spectrum across a range of frequencies. It makes it more difficult to identify the central frequency of a peak in the spectrum or obtain a correct estimate of its width. In many cases it is impossible to identify the peak at all. The best solution to the problem of spectral leakage is to measure a longer time sequence in the first place, but this is not always possible.

An example is given by Dahlen (1982) for a synthetic normal mode spectrum from an explosive source at a distance of 95°. The ideal spectrum is shown in Figure 3.3 (lowest trace). It was calculated theoretically and would be obtained from a record of infinite length, if such a record could be measured. The middle spectrum of Figure 3.3 corresponds to a boxcar window; the spectrum is the same as would be obtained from a 20 h record of the same ground motion. The peaks have been broadened by the finite length of record. Prominent side lobes appear as the fuzzy in-fill between the major peaks; they overlap throughout the low-frequency (long-period) part of the spectrum.

Shortening a record broadens spectral peaks and introduces side lobes, particularly at low frequencies corresponding to periods that are shorter than the total length of record by only a small factor. This is a fundamental limitation placed on the data by the limited recording time.

It is possible to reduce the side lobes by changing the window function, but at a price. Instead of using a boxcar window we can use some other function with a more favourable Fourier transform, one that has smaller or no side lobes. One example is the triangular or Bartlett window, in which the time sequence is multiplied by a function like

$$\begin{aligned} W(t) &= t \quad 0 \le t \le T/2 \\ &= (T - t) \quad T/2 \le t \le T. \end{aligned} \tag{3.8}$$

Note that this is the result of convolving a boxcar function with itself. The Fourier transform is therefore the square of that of the boxcar in Figure 2.2. The side lobes are still present but are less of a problem because the central peak has been increased at their expense. As a general rule, side lobes are reduced by smoothing the sharp edges of the rectangular boxcar window. This is called *tapering*.

An extreme case is the Gaussian window, which has no side lobes. The integral Fourier transform of the function $\exp\left(-t^2/\tau^2\right)$ is proportional to $\exp\left(-\tau^2\omega^2/4\right)$ (Appendix 2), which decreases monotonically as $\omega \to \pm\infty$. τ determines the width of the Gaussian; a small value of τ indicates a narrow function in the time domain but a broad function in the frequency domain, and vice versa. This is a general property of the Fourier transform, sometimes called the uncertainty principle (see Box 3.1). Ideally, for resolving spectral peaks, we want a narrow function in the frequency domain and therefore a broad function in the time domain, but its breadth is again limited by the length of the time series.

Gaussian windows are used in time–frequency analysis of seismograms (Dziewonski *et al.*, 1969), in which one seeks to identify the change in frequency with time of a dispersed wavetrain. The original seismogram is divided up into intervals and the spectrum is found for each interval. It is then possible to contour the energy in the wavetrain as a function of the central time of each window and frequency. For surface waves, the central time is proportional to the inverse of the group velocity of the waves. The Gaussian window is the obvious choice because it has the same form in both time and frequency domains; it puts both axes of the contour plot on an equal footing.

Box 3.1. The uncertainty principle

Heisenberg's uncertainty principle states the impossibility of knowing accurately both position and momentum of a particle; one can improve the accuracy of the determination of one of these two quantities only at the expense of the accuracy of the other. Position and momentum of, for example, an electron are described by the wave function or probability density functions (pdfs). Mathematically, the pdfs of position and momentum are Fourier integral transforms of each other. In the simple case when the position is confined by a Gaussian pdf, the momentum is confined by the transform, whose width is inversely proportional to the width of the position's pdf. This is the theory underlying the rough statement that the product of the uncertainties in position and momentum is constant.

In time series analysis the uncertainty principle gives a compromise between the accuracies of the height and location (or central frequency) of a spectral peak. We can improve on the accuracy of the location of the peak by making the peak narrower, but if we do so we introduce a larger error in the spectrum, which makes the height unreliable. Similarly, we can smooth the spectrum to determine the height of a peak more precisely, but this spreads the peak and makes the determination of its central frequency less accurate. The product of the errors remains roughly constant, and would be exactly constant if the peak had the Gaussian bell shape.

So far we have only been concerned with resolution of the spectrum, the need to separate neighbouring spectral peaks, or to be sure that a high point in a continuous spectrum is within a measured frequency range. There is also the problem of unwanted noise in the original time sequence, which will map faithfully into the Fourier transform through the DFT formula (2.17). Convolution with a broad function is a smoothing operation, and can act to reduce noise when it dominates the rapidly varying part of the spectrum. Time-windowing can therefore smooth the spectrum and reduce noise, but at the expense of spectral resolution. This is yet another instance of a recurring trade-off between noise reduction and resolution in data processing.

A whole menagerie of windows with different statistical properties has been used in geophysics. All windows suffer the same unavoidable problem: reducing

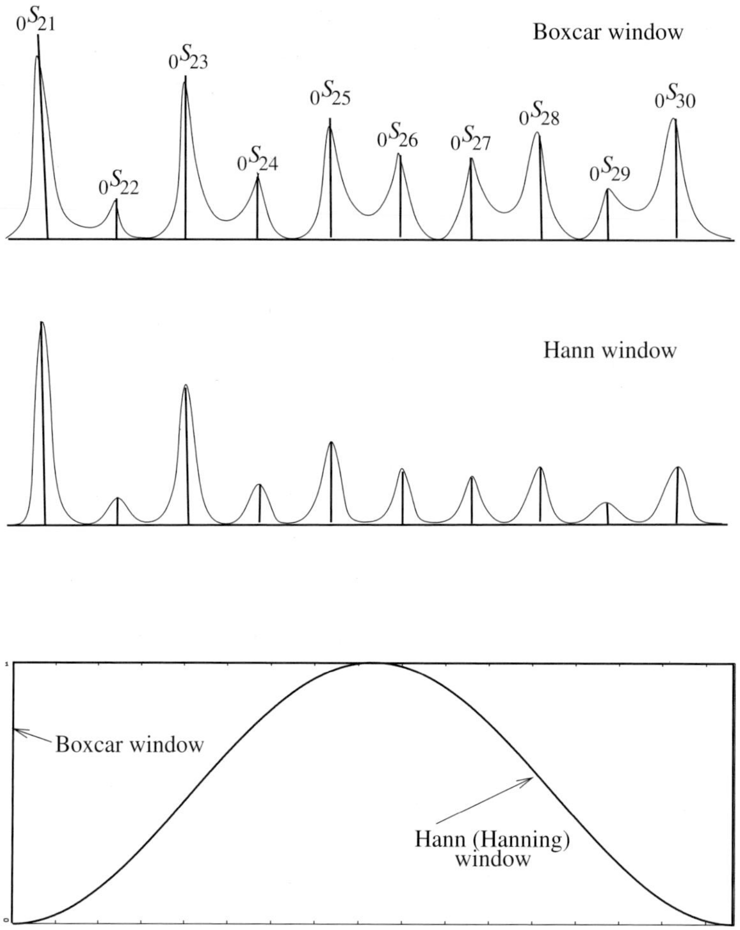

Fig. 3.4. Detail of effect of Hann taper on spectral peaks, and shape of Hann window (lower figure). Reproduced, with permission, from Dahlen (1982).

side lobes broadens the central peak. The choice of window is determined to some extent by what we want to measure. If the peaks are clearly separated and easily identifiable then side lobes are not a problem: a boxcar window will produce a good narrow peak. If the noise is large a broader window will be needed to separate them.

Returning to Figure 3.3, the top trace shows the spectrum for the same idealised time series using a 20 h long Hann window. This spectrum could be obtained from the same length of record as the middle trace (boxcar taper). The spectral peaks are broader but the side lobes are much reduced. The effect is shown in more detail in Figure 3.4; note that the boxcar has the greater spectral leakage (in-fill between the peaks) and narrower central peaks.

Other popular windows are Blackman–Harris, also studied by Dahlen (1982), and the cosine taper, in which the boxcar is tapered at each end by a half cycle of a cosine. I find this approach to tapering unsatisfactory: it is arbitrary and involves a loss of data by downgrading data away from the centre of the record. Modern optimised multi-taper methods avoid this problem. They are discussed in Section 10.3.

Box 3.2 summarises the methods used to digitise a continuous time series and produce a spectrum.

Box 3.2. To digitise a continuous time series and produce a spectrum

(i) *Filter* the analogue record to remove all energy near and above the Nyquist frequency. The signal is now band-limited.
(ii) *Digitise* using a sampling interval such that the Nyquist frequency lies above the highest frequency in the original data. The sampling theorem ensures that no information is lost.
(iii) *Window* the data with a finite-length window. In practice this may be imposed by the duration of the experiment, or you may be choosing a suitable piece of a continuous recording (e.g. a fixed time of a continuous seismometer recording starting at the origin time of the earthquake).
(iv) *Detrend* (e.g. by removing a best-fitting straight line). Slow variations across the entire time sequence contain significant energy at all frequencies and can mask the high frequency oscillations of interest.
(v) *Taper* to smooth the ends of the time series to zero. This reduces spectral leakage but broadens any central peaks. It also has the effect of equalising end points of the data to zero, thus avoiding any discontinuities seen by the DFT from periodic repetition. Tapering is equivalent to convolution in the frequency domain; it smooths the spectrum.
(vi) *Pad with zeroes*, usually to make the number of samples a power of 2 for the FFT. This reduces the spacing in the frequency domain and smooths the spectrum by interpolation.

3.3 Examples of spectra

3.3.1 Variations in the Earth's magnetic field

The Earth's magnetic field is influenced by the Sun and its position relative to the local position on the Earth's surface. These changes are brought about by a combination of factors, including ionisation of the upper atmosphere, which changes the electrical conductivity, and movement of the ring current within the magnetosphere by the angle of incidence of the solar wind on the Earth's main dipolar

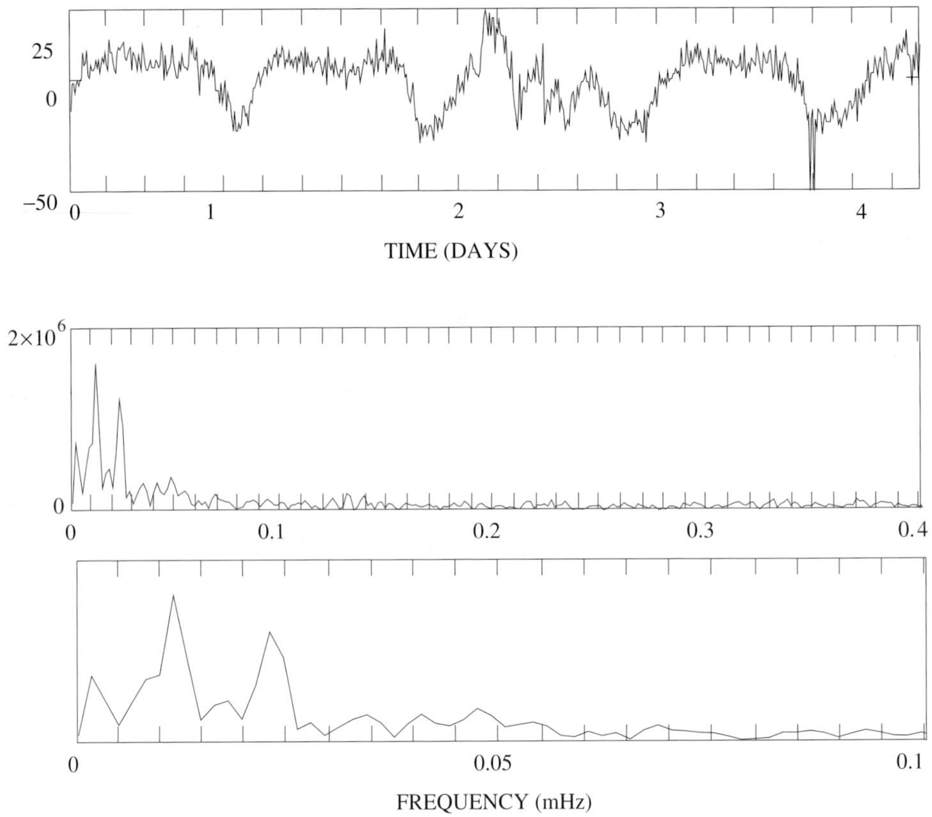

Fig. 3.5. Proton magnetometer data. The top panel shows the raw data, the middle panel shows the spectrum, and the lower panel shows the low-frequency portion of the spectrum with diurnal and semi-diurnal peaks.

field. The changes range from a few nanotesla (nT) during quiet days to many thousands of nT during a severe magnetic storm and are easily detectable by a proton magnetometer.

The time sequence plotted in Figure 3.5 was obtained by running a proton magnetometer for several days to record the total intensity of the Earth's magnetic field every 10 min. The proton magnetometer is a true absolute, digital instrument. It works on the principle that a proton possesses its own magnetic moment, which is known from laboratory experiment. The proton precesses in a magnetic field and the precession frequency is proportional to the field strength. The amplitude of the precession gradually decays until the proton's moment is aligned with the applied field. The instrument aligns protons with an applied field which is then removed; the protons then precess under the influence of the Earth's magnetic field until they realign with it. Each precession is counted and the total number divided by the time to give the precession frequency and hence the magnetic field strength. The instrument is digital because it counts an integer number of precession cycles, and is absolute because the result depends only on knowing the magnetic moment of the proton: the instrument needs no calibration.

The magnetometer was coupled to a simple recorder that printed out the time and field strength every 10 min. The data were recorded on floppy disk. The raw data and amplitude spectrum are shown in Figure 3.5. The spectrum was calculated by first removing the mean value of the data (otherwise the spectrum is completely dominated by the zero-frequency value and it is impossible to see anything in the plot), padding to 1024 points, and multiplying by a cosine taper with the cosine occupying 10% of the full length of the time series. Finally, only the low-frequency part of the spectrum is plotted to show the two interesting peaks.

The sampling interval is 10 min and sampling frequency 1.667 mHz. The Nyquist frequency is half the sampling frequency, or 0.8333 mHz. The sequence contains 625 samples, a total duration of 4.34 days. The sequence was padded to 1024 values, so the frequency interval in the spectrum is 1.623 μHz. There are two peaks in the spectrum, one at about 11.5 μHz and one at about twice that frequency. These correspond to periods of 24 and 12 h respectively. They are probably an effect of the Sun on the ionosphere. Oscillations are apparent in the raw time series in Figure 3.5, but one could not determine the presence of two frequencies by inspecting the time sequence alone.

3.3.2 The airgun source

We do not know the time function for an earthquake source in advance; we need to determine it from the available data, and it is often difficult or impossible to separate the effects of a complicated source and a complicated Earth structure on the seismogram. Explosion seismology is different, because it is often possible to run separate experiments to measure the source function. The ideal source is a delta function in time, since this would convolve with the ground structure to produce a seismogram that reflects the structure exactly, but such a source is impossible to realise physically.

A time function for an airgun source in water and its spectra are shown on the left of Figure 3.6. The initial pulse is followed closely by a reflection of opposite sign from the water surface, the lag being determined by the depth of the airgun beneath the water surface. This double pulse leads to interference at certain frequencies ν, when the travel times of the direct and reflected waves differ by $(n + \frac{1}{2})/\nu$, where n is an integer. These frequencies correspond to the zeroes in the power spectra, which can be seen in Figure 3.6 to vary with the depth of the source.

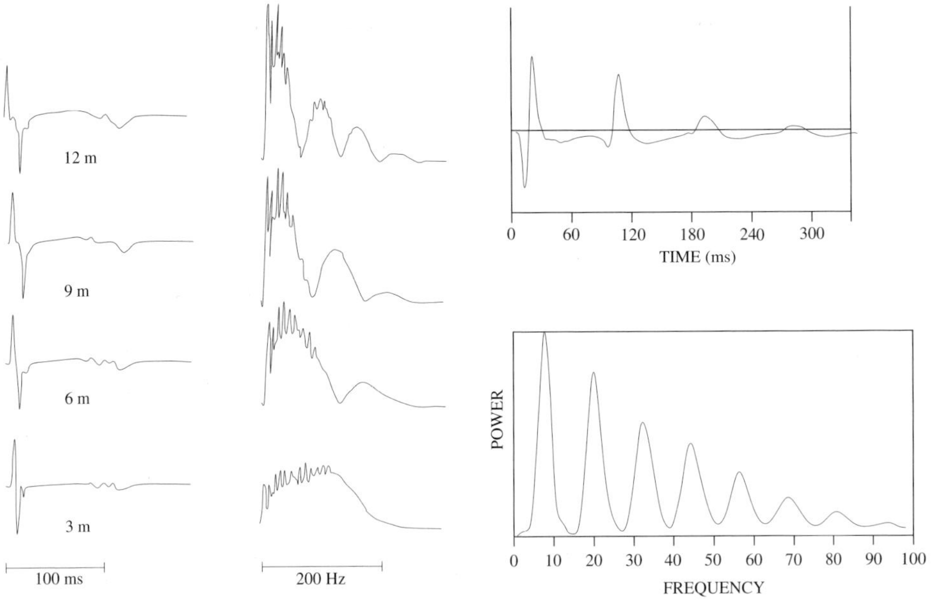

Fig. 3.6. Variation of the time function of an airgun source as a function of depth below water surface, and power spectra. Reproduced, with permission, from BP.

The airgun also releases a bubble of air that oscillates, producing secondary pulses whenever the pressure reaches a maximum or minimum. The pulses occur at an interval determined by the natural frequency of oscillation of the bubble, which in turn depends on the size and type of the airgun. The trace on the top right of Figure 3.6 is for a 70 cubic inch gun. The bubble pulse imparts a long characteristic time (here about 100 ms) to the time sequence, which maps into low-frequency structure (10 Hz) in the power spectrum (bottom right).

3.3.3 Microseismic noise

It is common to conduct a noise survey in both earthquake and explosion seismology by simply recording incoming energy in the absence of any source. Figure 3.7 shows spectra from a broadband seismometer operating in New Zealand, where microseisms can be large. Each curve represents the power spectrum of 15 min samples from the Z component taken every 3 h. The sampling rate was 25 Hz. Note the broad peaks at periods of 6 s and 3–4 s typical of microseisms. Removing microseismic noise is an important stage in processing modern broadband seismograms.

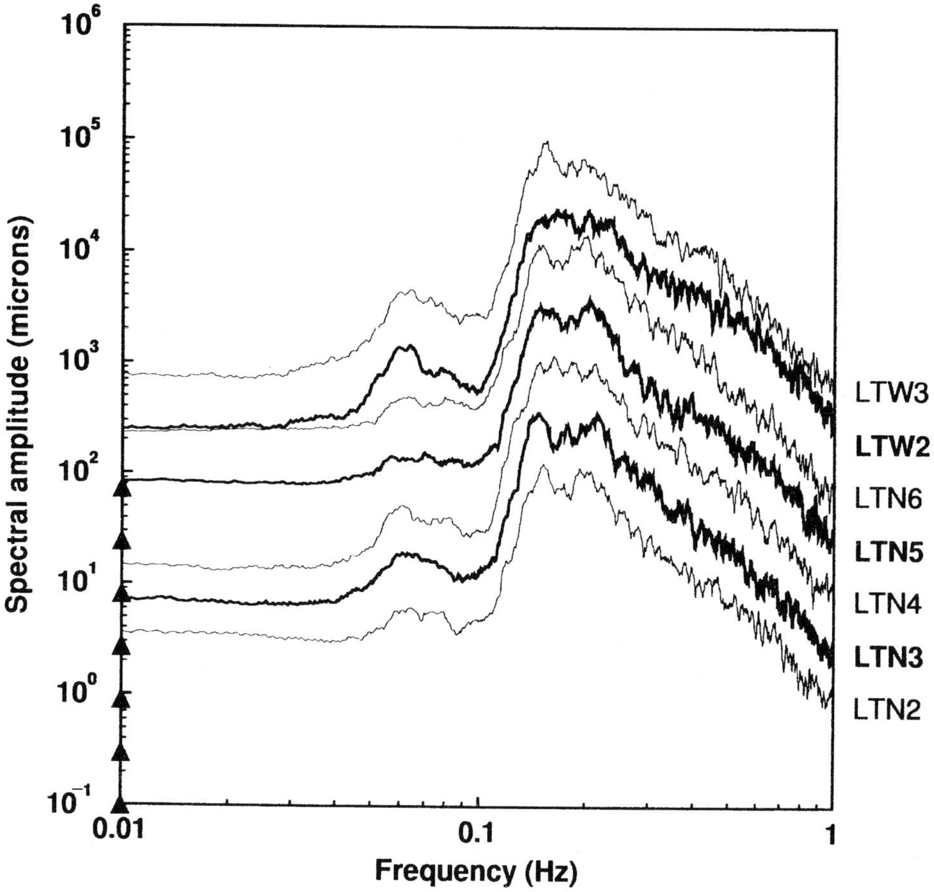

Fig. 3.7. Power spectra of microseismic noise from a broadband seismometer in New Zealand. Each curve is derived from 15 min of data at intervals of 3 h. Each curve has been offset by the amount indicated by the arrows on the y-axis for clarity. Reproduced, with permission, from Stuart *et al.* (1994).

EXERCISES

3.1. You decimate a seismogram from 25 Hz sampling to 1 Hz sampling in order to reduce the volume of data required in a study of surface waves with 20 s period. Give the old and new Nyquist frequencies. What previously meaningful periods have been aliased into 20 s period?

3.2. A time series is sampled at 250 Hz. What are the sampling interval and Nyquist frequency? The original signal, before digitising, was contaminated with noise at about 400 Hz. At what frequency would this noise appear in the spectrum of the digitised sequence? Repeat the calculation for noise contamination at 475 Hz.

3.3. You hope to find two peaks in the spectrum of a gravimeter record. Theory predicts they will have periods of 6.2 h and 6.6 h with good signal to noise ratio.

(a) For how long must you record in order to resolve two distinct peaks?
(b) What is a suitable sampling frequency, assuming no significant energy arrives with period <6 h?
(c) How many values will your time series contain?
(d) What else would you do to obtain a practical estimate of the peak frequencies?

3.4. Sketch the spectrum you would expect from seismograms recorded under the following circumstances. Label the axes.
A noise study reveals a generally white background of amplitude 3×10^{-6} m s^{-1}, plus microseismic noise of amplitude 10^{-3} m s^{-1} in the period range 4–11 s. Teleseismic body waves have typical amplitude 3×10^{-4} m s^{-1} and frequency 0.5–10 Hz; surface waves have typical amplitude 5×10^{-2} m s^{-1} and period 10–100 s.

3.5. **Computer exercise: aliasing**
Use the `pitsa` program to generate the same sinusoidal time series as in the previous practical 2.11, but now include frequencies 10, 11, 12, 13, 14, 15 Hz and higher if you like. Take the amplitude spectra.

(a) Explain the spectrum at 11 Hz. Repeat at higher frequencies.
(b) What happens to the spectrum at 10 Hz? Try comparing 10 Hz signals with phases 0° and 90°.
(c) You may have noticed that some of the high-frequency sine waves in this and the previous computer exercise look strange when displayed on the computer screen. This is also related to the problem of aliasing – can you explain it?

3.6. **Computer exercise: spectral resolution, tapering and side lobes**

(a) Take the Fourier transform of a boxcar function as follows
`Utilities → Test Signals → Constant`
`Advanced Tools → Spectrum → FFT → Boxcar (taper fraction 0.5)`
Note the shape of the boxcar function before you display the amplitude spectrum. You can blow up part of the spectrum display with
`Utilities → Header access → plot → viewport`
and selecting the appropriate x- and y-axis limits. Compare the result with the analytical formula from lectures.
(b) Generate a sine wave with a frequency of 0.625 Hz, 20 Hz sampling rate and 51.2 s length. Use the boxcar taper to decrease the effective length of this time series by a half, a quarter, etc., while padding the oscillation with zeroes such that the length of the padded time series remains 51.2 s. This can be easily achieved by setting the `taper fraction` to 0.5, 0.75, etc. Calculate the spectra. What do they tell you about resolution?
(c) Calculate the spectra of the tapered oscillations in the previous question without padding with zeroes by changing the number of points in the FFT, which must

always be a power of 2. Compare the results with the previous question and repeat the process with a time series of a slightly different frequency.

(d) Generate a signal that consists of two decaying sine waves of slightly different frequencies. Use `Utilities` → `Test signals` → `Mixed Signal` to generate a decaying sine wave (e.g. sample frequency 1 Hz, 1024 points, signal frequency 0.05 Hz (20 s) and exponential decay factor 0.01). Generate a second, similar signal with frequency 0.04 Hz (for example). Add them together with
`Utilities` → `Trace Utilities` → `Double Channel Utilities` → `aX + bY`.
Lastly take the spectrum with boxcar taper, taper fraction 0. This is the best you can do with exponentially decaying series.

(e) Generate similar noisy signals and take the spectrum again. In order to simulate some problems in estimating the spectral peaks, vary the length of the time series.

Use one of your noisy time series and take the spectrum again, this time using rather short tapers, first the boxcar, then the Hanning, then other tapers provided by `pitsa`. Compare their impact on spectral resolution.

3.7. **Computer exercise: normal modes**
Read the ASCII file `cmo.x`. ASCII is a standard code to translate characters into binary numbers. It is used on most computers. You can read an ASCII file with an ordinary text editor (try it); you cannot read files with special formats such as ISAM (try it). Each file has a *header* containing any information you might need to understand the time sequence. The time sequence itself follows the header. For example, the header might tell you the sampling rate, date of acquisition, or instrument characteristics. The header for this file is 30 lines long. We shall ignore the header information so enter 30 when `pitsa` asks how how many lines to skip. The sampling interval is 20 s (i.e. the sampling frequency is 0.05 Hz or 50 mHz). The recording is from a gravimeter (acting here as a very long-period seismometer) at College, Alaska, of a very large, shallow, earthquake in Oaxaca, Mexico.

(a) The surface waves were so large, the instrument was saturated and took about 2 h to recover. Note the nonlinear behaviour before it settles down. `ZOOM` out the bad data at the beginning of the record with
`Routine Tools` → `Zoom` → `Untapered`
and choose the `Double cursor` option. Can you see the Earth tides? Remove them with a running average of 6 h using
`Routine Tools` → `Baseline correction` → `Running average (remove)`
(It is better to calculate them from theory and subtract them, but `pitsa` cannot do this.) Take the amplitude spectrum using any taper you like. Experiment with different tapers and taper fractions in `pitsa`'s menu.

(b) Identify the peaks in the frequency range 1.9–3.0 mHz and measure their frequencies to 0.01 mHz (you will need to plot the spectrum on a finer scale for this). These are the fundamental spheroidal normal modes of the Earth. They are labelled, by convention, ${}_0S_l$ where the first zero denotes a fundamental (overtones have this 'quantum number' greater than zero) and l is the 'angular order number', giving essentially the number of wavelengths of the mode (or standing wave) between North and South Poles. Thus l is a sort of wavenumber. Compare your results with the calculated values for the Earth

$$
\begin{array}{ll}
{}_0S_{12} & 1.98917 \\
{}_0S_{13} & 2.11170 \\
{}_0S_{14} & 2.23017 \\
{}_0S_{15} & 2.34521 \\
{}_0S_{16} & 2.45712 \\
{}_0S_{17} & 2.56612 \\
{}_0S_{18} & 2.67241 \\
{}_0S_{19} & 2.77617 \\
{}_0S_{20} & 2.87764
\end{array}
$$

and identify the modes.

(c) Repeat the procedure with the file `esk.x`, which is for the same earthquake recorded at Eskdalemuir, in Scotland. There are glitches towards the end of the record; these can be removed with

`Routine Tools` → `Edit` → `glitch editing`

or you can just `ZOOM` out the last part of the record (this wastes data).

4

Processing of time sequences

4.1 Filtering

Filtering of a time sequence is convolution with a second, usually shorter, sequence. Many important linear processes take the form of a convolution; for example the seismometer convolves the ground motion with its own impulse response, and can therefore be thought of as a filtering operation. We shall concentrate on *bandpass* filters, which are designed to eliminate whole ranges of frequencies from the signal. Filters that eliminate all frequencies above a certain value are called *low-pass* while those eliminating all frequencies below a certain value are called *high-pass*. A *notch* filter removes all frequencies within a certain band. The range of frequencies allowed through is called the *pass band* and the critical frequencies are called *cut-off* frequencies. These filters are particularly useful when signal and noise have different frequency characteristics.

An example of filtering is shown in Figure 4.1. Exactly the same seismogram shown in Figure 1.1 has been high-pass filtered to remove energy at frequencies below 1 Hz. This removes large surface waves to reveal body waves. Surface waves have typical periods of 20 s while body waves have much shorter periods of about 1 s, so they are quite widely separated in frequency. An aftershock is concealed in the surface waves generated by the main event. A high-pass filter does a good job of removing surface waves from the record, leaving the body waves from the aftershock unchanged.

Consider first a moving average, in which each element of the time sequence is replaced by an average of M neighbouring elements. This is a convolution of the original time sequence with a boxcar function, and is therefore a filtering operation. We might expect the averaging process to smooth out the rapid variations in the sequence and therefore act as a low-pass filter. To examine the effect properly we must examine the operation in the frequency domain. Convolution in the time domain is multiplication in the frequency domain, so we have

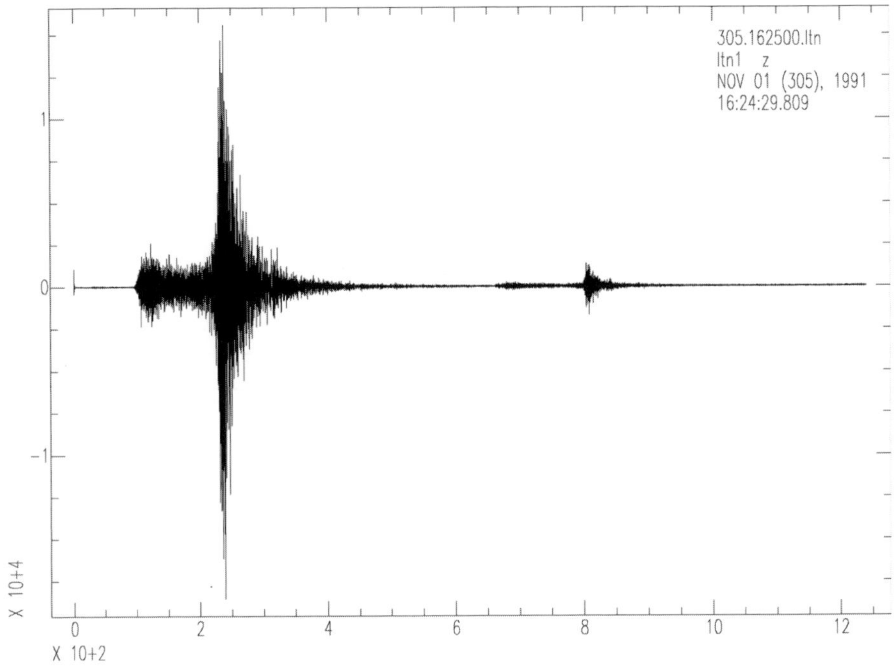

Fig. 4.1. The same data as displayed in Figure 1.1, but here Butterworth high-pass filtered with $\omega_c = 1$ Hz and four poles. The surface waves have been completely removed and an aftershock, which had previously been obscured by the surface waves, can be seen.

multiplied by the Fourier transform of the boxcar (Figure 2.3). Its shape is nowhere near that of the ideal low-pass filter because of the side lobes and narrow central peak. The amplitude spectrum is identically zero at the frequencies nN/MT: these correspond to sine waves with exactly n cycles within the boxcar. The moving average is therefore a very good way to remove unwanted energy with an exact known frequency (a diurnal signal for example) but is not a good general low-pass filter because it lets through a lot of energy with frequency above the intended cut-off.

An ideal low-pass filter should have an amplitude spectrum that is equal to zero outside the pass band and equal to unity inside it. The sharp discontinuity cannot be represented well by a finite number of terms in the time domain because of Gibbs's phenomenon; a compromise is needed between the sharpness of the cut-off and the length of the sequence in the time domain. The compromise is similar to that in tapering windows for spectra. A large variety of tapered cut-offs are in use, but we shall concentrate on the Butterworth, which for low-pass filters has the functional form

$$|F_l(\omega)|^2 = \frac{1}{1 + (\omega/\omega_c)^{2n}}, \tag{4.1}$$

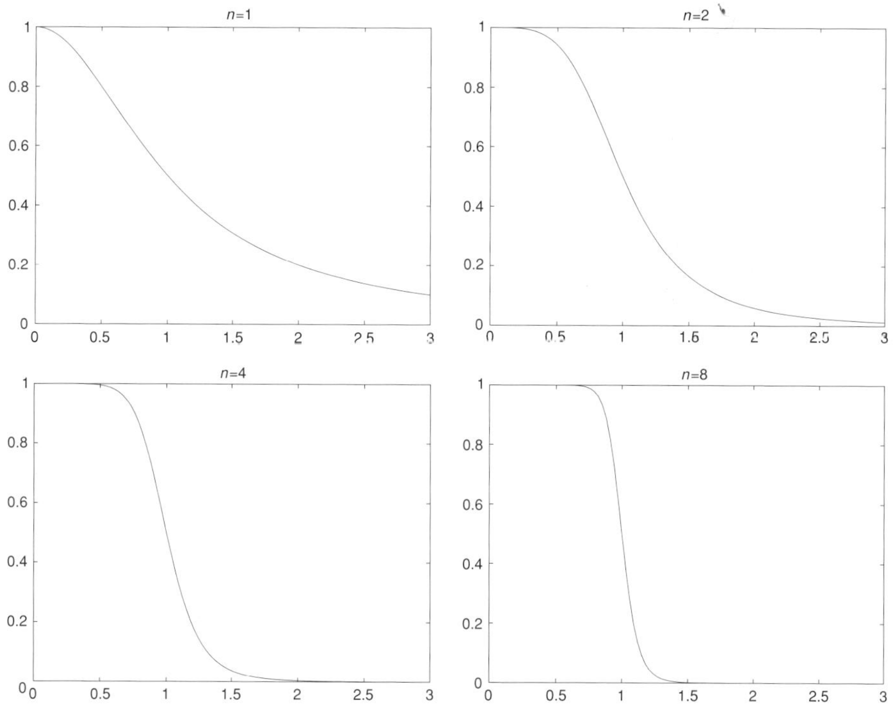

Fig. 4.2. Variation with n of the amplitude spectra defined by the function given in equation (4.1). The cut-off frequency is kept at 1 Hz; note the sharper roll-off as n is increased.

where ω_c is the cut-off frequency at which the energy is reduced by half, and the index n controls the sharpness of the cut-off. The n is the number of *poles* of the amplitude spectrum. Examples of the shape of this function are shown in Figure 4.2. Figure 4.3 shows the effect of filtering a seismogram with the same cut-off frequency and different numbers of poles.

Other filters may be constructed from the low-pass function $F_\ell(\omega)$. A high-pass filter has amplitude spectrum $F_h(\omega) = 1 - F_\ell(\omega)$. A bandpass filter may therefore be constructed by shifting the same function along the frequency axis to centre it about ω_b

$$|F_b(\omega)|^2 = \frac{1}{1 + [(\omega - \omega_b)/\omega_c]^{2n}} \tag{4.2}$$

and a notch filter by $F_n(\omega) = 1 - F_b(\omega)$. $F_\ell(\omega)$ is symmetric about $\omega = 0$, because the corresponding time sequence is real.

To construct the time sequence we must also specify the phase, then take the inverse Fourier transform. The simplest choice is to make the phase zero and simply invert the amplitude spectrum. This gives a *zero-phase* filter. Zero-phase filters obviously do not change the phase of the original time sequence, which also means

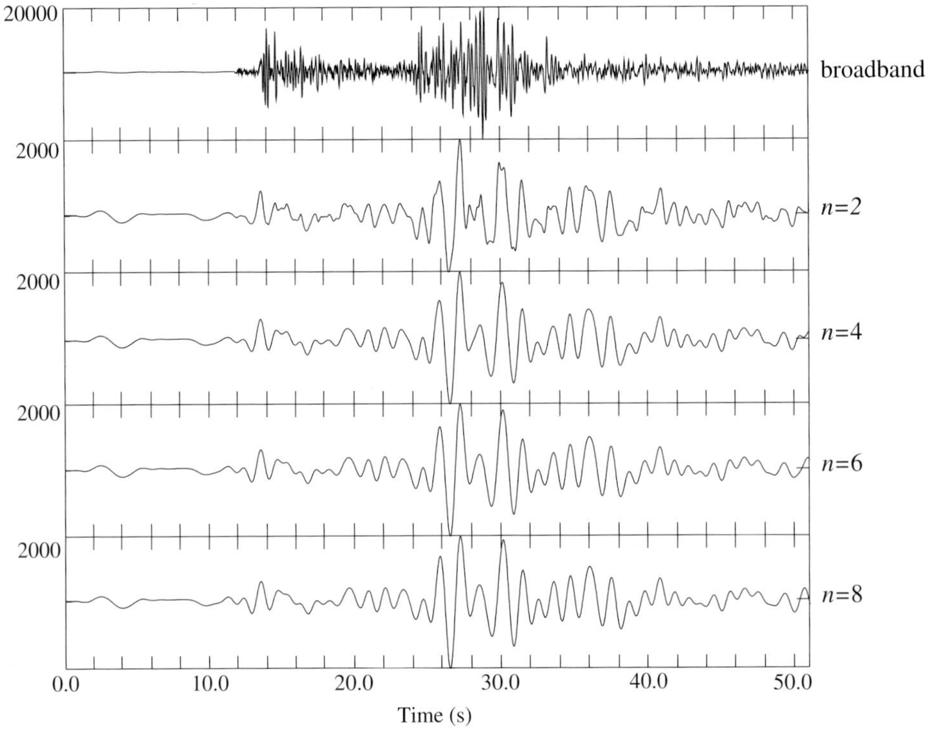

Fig. 4.3. A broadband seismogram from a regional event which ocurred on 20 February 1992 at 40.818° S, 174.426° E, depth 74 km, magnitude $m_b = 4.0$, and was recorded in New Zealand. Butterworth low-pass filtered with the same cut-off frequency (1 Hz) and varying number of poles (n). P and S wavetrains are visible. Note the further reduction in high-frequency content as the number of poles is increased.

that they will not in general shift peaks of the original signal. The time sequence of a zero-phase filter must be symmetrical about the origin because the Fourier transform is real. The filter will therefore be *acausal*: the output of the filter at time t depends on input values earlier than time t. It is not possible to construct a physical device to perform such a filtering operation, but it is quite easy when processing a time sequence after recording. Such filters are also called *non-realisable*.

An easy way to produce a zero-phase filter in the time domain is to convolve first with the chosen filter, then time-reverse it and convolve again. The first operation multiplies by $F(\omega)$ and the second by $F^*(\omega)$ because time reversal complex conjugates the Fourier transform (Section 2.3.3), giving a net multiplication by $|F(\omega)|^2$, which is real. These are called *two-pass* filters.

A *causal* or *realisable* filter is defined as one producing output for time t that depends only on input values from time t or later. It cannot be zero phase. An infinite number of causal filters can be produced from a single amplitude spectrum,

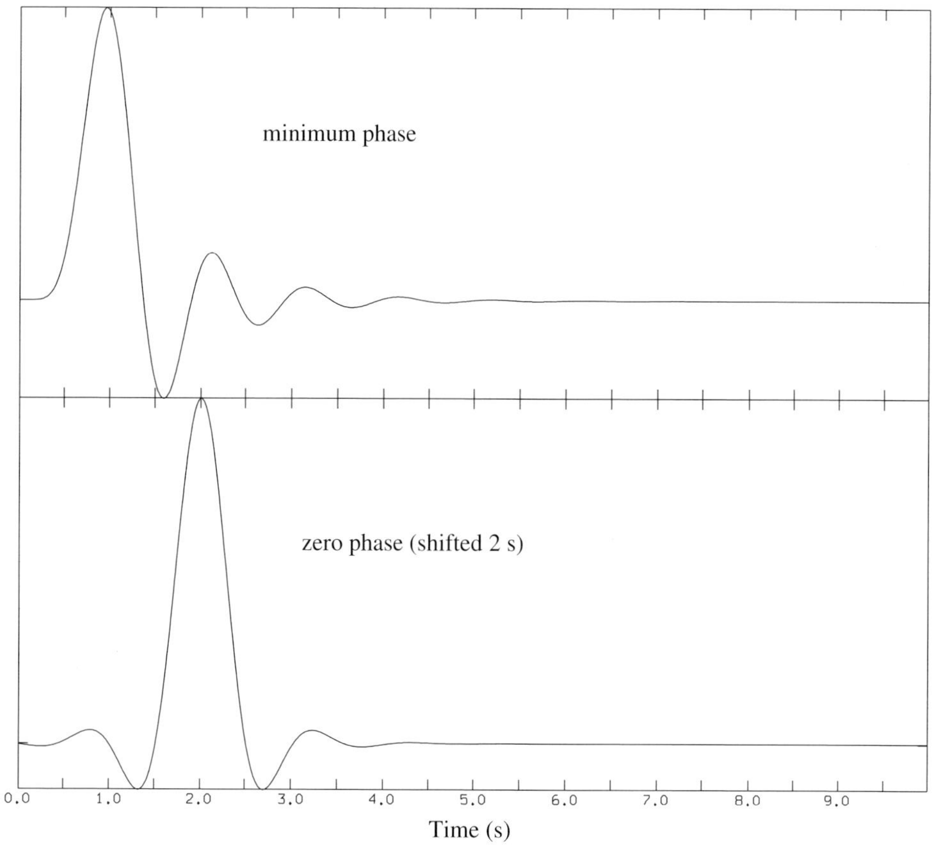

Fig. 4.4. Two filters in the time domain with the same amplitude spectrum given by equation (4.1) with $n = 4$. The upper filter is causal and minimum phase. The lower filter is zero phase and should be symmetric about $t = 0$: it has been shifted by 2.0 s to show the shape in relation to the minimum phase filter. It is acausal.

depending on the choice of phase. The most useful choice is *minimum phase*, giving a time sequence with energy concentrated at the front. A minimum-phase filter has the desirable property of being the 'shortest' filter with the given amplitude spectrum. Examples of zero-phase and causal minimum-phase filters with the same amplitude spectrum are shown in Figure 4.4.

The minimum-phase or *minimum-delay* filter may be constructed by factorising the z-transform:

$$F(z) = f_{N-1} \prod_{i=0}^{N-1} (z - z_i) . \tag{4.3}$$

Each factor $(z - z_i)$ has a phase $\phi_i(z) > 0$ and the phase of $F(z)$ will be their sum, which is minimised by minimising the phase of each factor while retaining the same amplitude. A general factor $a + bz$, with $z = x + iy$, has squared amplitude $a^2 + 2abx + b^2$; keeping this constant for all x requires fixing both $a^2 + b^2$ and ab, which is only possible for the combination $\pm(a, b), \pm(b, a)$. The minus signs are taken up as a π phase change in making the phase positive, leaving only the alternatives (a, b) and (b, a). We can therefore change a maximum phase factor with $b > a$ to a minimum phase factor by swapping a and b. Each factor of the z-transform (4.3) is therefore either minimum or maximum phase, depending on whether $z_i > 1$ or $z_i < 1$. The whole z-transform is minimum phase only if all its constituent factors are minimum phase. We can therefore form a minimum delay sequence by changing all the maximum phase factors from $z - z_i$ to $1 - z_i z$. The amplitude spectrum is unchanged. Note that the z-transform of a minimum delay time sequence has all its zeroes outside the unit circle. This method of finding a minimum-delay time sequence is impractical for long sequences because of numerical difficulties associated with finding roots of high order polynomials; a more practical method based on Kolmogoroff spectral factorisation will be described in Section 4.3.

We now show that a minimum-phase sequence has all its energy concentrated at the front. This involves proving that all the partial sums of the squares of terms in a minimum-phase sequence F:

$$E_p^f = \sum_{k=0}^{p-1} |f_k|^2 \tag{4.4}$$

are greater than or equal to the corresponding partial sums for any other sequence with the same amplitude spectrum.

We generalise to complex sequences and introduce a second sequence g that differs from f by only one factor in the z-transform:

$$F(z) = (z - z_0)\, C(z) \tag{4.5}$$

$$G(z) = (-z_0 z + 1)\, C(z), \tag{4.6}$$

where

$$C(z) = \sum_{k=0}^{K-1} c_k z^k. \tag{4.7}$$

Note that f and g have the same amplitude spectrum, and that $|z_0| > 1$ for f to be minimum delay.

The $(2k)$th term in the expansion of the difference $E_p^f - E_p^g$ is then $(|z_0|^2 - 1)$ $(|c_k|^2 - |c_{k-1}|^2)$. Summing from $k = 0$ to $p - 1$ gives

$$E_p^f - E_p^g = (|z_0|^2 - 1)\left|c_p\right|^2 > 0. \tag{4.8}$$

The f sequence therefore has more energy at the front. The same follows for all factors of $F(z)$ and therefore the minimum phase sequence has more energy concentrated at the front that any other sequence with the same amplitude spectrum. This result, due to E. A. Robinson, is called the *energy-delay theorem*.

4.2 Correlation

The cross correlation of two time sequences a and b is defined as

$$c_k = \frac{1}{N+M-1} \sum_p a_p b_{k+p}, \tag{4.9}$$

where k is called the *lag*, N and M are the lengths of a and b respectively, and the sum is taken over all $N + M + 1$ possible products. The *autocorrelation* was defined in (2.44); it is now seen as the cross correlation of a time sequence with itself

$$\phi_k = \frac{1}{2N-1} \sum_p a_p a_{k+p}. \tag{4.10}$$

The *correlation coefficient* is equal to the cross correlation normalised to give unity when the two sequences are identical and therefore perfectly correlated

$$\psi_k = \frac{\sum_p a_p b_{k+p}}{\sqrt{\sum_p a_p a_p \sum_p b_p b_p}}. \tag{4.11}$$

Note that $|\psi_k| \leq 1$, and that $\psi_k = 1$ occurs only when $a = b$ (perfect correlation). $\psi_k = -1$ when $a = -b$ (perfect anticorrelation).

As an example, consider the two sequences:

$$a: \quad 1, 2, 3, 2, 1; \qquad (N = 5) \tag{4.12}$$

$$b: \quad 1, -1, 1, -1, 2 \qquad (M = 5)\,. \tag{4.13}$$

The cross correlation c of a and b is of length $N + M - 1 = 9$:

$$\begin{aligned}
9c_{-4} &= a_4 b_0 = 1 \\
9c_{-3} &= a_3 b_0 + a_4 b_1 = 1 \\
9c_{-2} &= a_2 b_0 + a_3 b_1 + a_4 b_2 = 2 \\
9c_{-1} &= a_1 b_0 + a_2 b_1 + a_3 b_2 + a_4 b_3 = 0 \\
9c_0 &= a_0 b_0 + a_1 b_1 + a_2 b_2 + a_3 b_3 + a_4 b_4 = 2 \\
9c_1 &= a_0 b_1 + a_1 b_2 + a_2 b_3 + a_3 b_4 = 2 \\
9c_2 &= a_0 b_2 + a_1 b_3 + a_2 b_4 = 5 \\
9c_3 &= a_0 b_3 + a_1 b_4 = 3 \\
9c_4 &= a_0 b_4 = 2
\end{aligned}$$

giving the time sequence:

$$c = (1, 1, 2, 0, 2, 2, 5, 3, 2)/9. \tag{4.14}$$

Similarly, the autocorrelation of a may be evaluated as

$$\phi_a = (1, 4, 10, 16, 19, 16, 10, 4, 1)/9 \tag{4.15}$$

and for b

$$\phi_b = (2, -3, 4, -5, 8, -5, 4, -3, 2)/9. \tag{4.16}$$

The correlation coefficients are:

$$\psi = (1, 1, 2, 0, 2, 2, 5, 3, 2)/\sqrt{136}. \tag{4.17}$$

Now consider correlation in the frequency domain with $N = M$. Equation (4.9) is actually convolution of b with the time-reversed sequence a':

$$\begin{aligned} c_k &= \sum_{p=0}^{N-1} a_p b_{k+p} = \sum_{p=0}^{N-1} a'_{-p} b_{k+p} \\ &= \sum_{p=0}^{-N+1} a'_p b_{k-p} = \sum_{p=0}^{N-1} a'_p b_{k-p}. \end{aligned} \tag{4.18}$$

The last step is possible if the sequences are periodic, because then the sum is over an entire period. Taking the DFT and using the convolution theorem (2.6) and the result that time reversal is equivalent to complex conjugation in the frequency domain (Section 2.3.3) gives:

$$C_n = A_n^* B_n. \tag{4.19}$$

It follows that the DFT of the autocorrelation is the power spectrum:

$$\Phi_n = |A_n|^2. \tag{4.20}$$

Φ_n is real and the autocorrelation is therefore symmetric in the time domain. Furthermore, a zero-phase, two-pass filter is the autocorrelation of the original filter.

A *matched filter* is used to cross correlate the data with an estimate of the known signal. It is particularly useful in seismology when the shape of the waveform is known but its arrival time is not. Suppose the time sequence comprises a deterministic signal with additive random noise that is uncorrelated with the signal:

$$a_k = s_k + n_k. \tag{4.21}$$

Cross correlating a with an estimate of s_k will reduce the random noise and leave the autocorrelation of the signal. The time reversal of s_k can be used to form the filter, so that convolution produces the cross correlation. Matched filters are good for searching a noisy time sequence for a signal of known shape. A typical example is to use a portion of the first P wave arrival, which is usually relatively clear, to make a filter to pick out later arrivals in the seismogram, such as reflections or refractions from deeper layers.

4.3 Deconvolution

Deconvolution is the reverse of convolution. It arises frequently in seismology, for example in removing or altering the instrument response of a seismometer, or calculating the response for a different source time function. There are four methods: recursion, as already given by equation (2.14) in Section 2.1, division in the frequency domain, construction of an inverse filter from the z-transform, and designing a suitable Wiener deconvolution filter by least squares, which will be dealt with in Section 10.2. All methods suffer from the same fundamental problems but it is instructive to compare all four.

4.3.1 Division in the frequency domain

Convolution is multiplication in the frequency domain, so deconvolution is simply division in the frequency domain. Suppose we wish to deconvolve sequence a from sequence c to give b; the frequency-domain procedure is to transform both sequences, perform the division

$$B(\omega) = \frac{C(\omega)}{A(\omega)}, \tag{4.22}$$

and inverse transform the result. A major problem arises when the amplitude spectrum of the denominator vanishes, at frequencies ω_0 where $|A(\omega_0)| = 0$, because we cannot divide by zero. Now if c is consistent with convolution of a with any other sequence, $C(\omega_0) = 0$ and equation (4.22) will yield $0/0$. However, the presence of small amounts of noise in either c or a will lead to wildly varying numerical estimates of $B(\omega)$ near ω_0. Inverse tranformation of these erroneous spectral values will map everywhere in the time domain, with disastrous consequences.

This difficulty is fundamental. Consider the common case of deconvolution of a seismometer instrument response: c is the seismogram and a the instrument impulse response. A short period instrument may give zero response for frequencies less than 0.5 Hz, so $A(\omega)$ is zero for $\omega < \pi$. There is no output from the

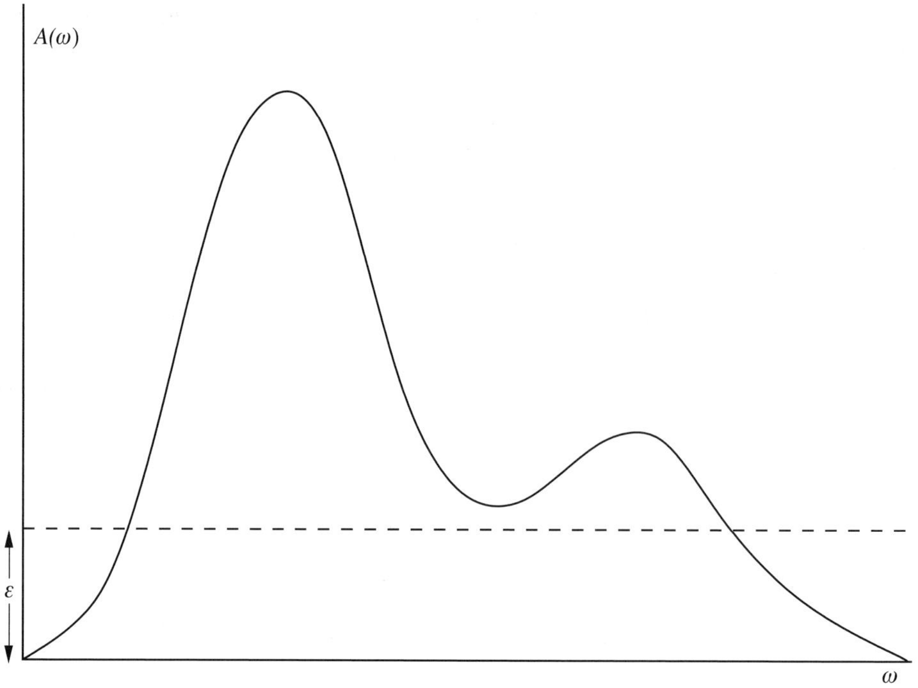

Fig. 4.5. Deconvolution involves division in the frequency domain. When the amplitude spectrum $A(\omega)$ falls to zero deconvolution is impossible, and when it falls below a certain value ε depending upon the noise, the result will be dominated by amplified noise. The problem is alleviated by adding a constant ε to fill in the holes in the spectrum, like filling them with water.

seismometer, yet by deconvolution we are trying to obtain the ground motion that was never measured in the first place!

We could design the deconvolution procedure to return zero whenever the instrument response, $A(\omega)$, is zero or falls below an assigned level. A similar procedure, which has similar effects, is the *water-level method.* A small constant ε is added to $|A(\omega)|$ (Figure 4.5). We are trying to estimate the ratio of two uncertain quantities, C and A, both of which contain noise denoted by σ_c and σ_a respectively: $(C + \sigma_c) / (A + \sigma_a)$. Provided $A \gg \sigma_a$ and $C \gg \sigma_c$ this ratio will be a good approximation to the required ratio C/A. However, at frequencies where $A(\omega)$ is comparable with σ_a the estimate of the ratio may become large. Adding a positive constant ε to A will *stabilise* this division, provided $\varepsilon \gg \sigma_a, \sigma_c$, because when $A = C = 0$ (as required for consistency) the ratio gives $\sigma_c / (\sigma_a + \varepsilon) \ll 1$. ε will also bias our estimate of the ratio C/A towards low values and change the phase, but these effects are acceptable provided ε is much less than typical values of $|A|$. Adding a constant to the spectrum is also referred to as 'adding white noise' to stabilise the deconvolution because the spectrum of white noise is flat (Section 2.4).

4.3.2 Constructing an inverse filter

Another way to view deconvolution is division by the z-transform, or multiplication by $1/A(z)$. If this inverse can be expanded as a power series in z we can extract the corresponding time sequence from the coefficients, which will give a causal filter provided there are no inverse powers of z. First factorise $A(z)$

$$A(z) = a_N \prod_{n=0}^{N} (z - z_n) \tag{4.23}$$

then expand the inverse by the method of partial fractions

$$\frac{1}{A(z)} = \sum_{i=0}^{N} \frac{\alpha_n}{(z - z_n)}. \tag{4.24}$$

We can expand each term in the sum as a power series. The general term gives

$$\frac{1}{(z - z_n)} = -\frac{1}{z_n}\left[1 + \frac{z}{z_n} + \left(\frac{z}{z_n}\right)^2 + \cdots\right]. \tag{4.25}$$

Combining all such power series and substituting into (4.24) gives a power series for $1/A(z)$. This power series will converge only if (4.25) converges for every term, which requires in general that

$$|z| < |z_n| \tag{4.26}$$

for every n. The Fourier transform is obtained by putting $z = \exp i\omega\Delta t$; hence $|z| = 1$ and (4.26) becomes $|z_n| > 1$ for all n. This is exactly the condition for a to be minimum phase. Therefore *only minimum phase sequences have causal inverse filters*.

At this point we digress to use this property to deduce an efficient method of finding a minimum-delay filter from a given amplitude spectrum. First note that the power spectrum $S(\omega)$ is obtained from the z-transform by setting $z = \exp i\omega\Delta t$ in $A(1/z)A(z)$. Taking the logarithm gives $\ln A(1/z) + \ln A(z)$. Inverting this z-transform produces causal terms from the second term while the first term yields only terms at negative times. Removing acausal terms in the time domain and taking the z-transform again gives $\ln A(z)$, then exponentiation gives the desired $A(z)$. The a obtained from inverting $A(z)$ will be causal because both mathematical functions ln and exp preserve causality: when applied to power series they both produce power series with only positive powers. It is minimum phase because the inverse filter is also causal: it can be produced by the same procedure starting from $S^{-1}(\omega)$. Problems arise when $S(\omega) = 0$, when neither $S^{-1}(\omega)$ nor $\ln S(\omega)$ exist, but these zeroes will always produce difficulties. This method is called *Kolmogoroff factorisation of the spectrum*. In practice the procedure is as follows: start with the power spectrum, take its logarithm, inverse transform, discard negative time terms, take the transform, and exponentiate.

Inverse filters have two potential advantages over division in the frequency domain: they are causal and, if they can be truncated, the convolution process may be more efficient than Fourier transformation of the entire input sequence. Truncation depends on how rapidly the power series for $1/A(z)$ converges, which depends ultimately on the size of the roots of the original z-transform; a root close to the unit circle will give a slowly convergent power series and a long inverse filter. The effects of these roots may be alleviated by a number of methods akin to the water level method, damping as in least-squares inversion (Section 10.2), or removal of the root altogether.

Damping is best illustrated by a simple example. Consider the first-difference filter

$$f = (1, -1) . \tag{4.27}$$

The z-transform of the inverse filter is $1/(1-z)$, which expands to

$$\frac{1}{1-z} = 1 + z + z^2 + \cdots \tag{4.28}$$

with radius of convergence 1 because of the root lying on the unit circle. The inverse filter is therefore of infinite length and of no practical use as it stands. Claerbout (1992) recommends replacing z with ρz, where ρ is slightly less than 1. Equation (4.28) then becomes

$$\frac{1}{1-\rho z} = 1 + \rho z + \rho^2 z^2 + \cdots , \tag{4.29}$$

which will converge quite rapidly even when ρ is only slightly smaller than 1. Claerbout (1992) calls this 'leaky integration': *integration* because it is the inverse of differencing, a discrete analogue of differentiation, and *leaky* because at each step a fraction $1 - \rho$ of the contribution to the integral is lost.

This fundamental problem with inverting the first-difference filter arises because it removes the mean value of the input sequence: this average can never be recovered. The zero-frequency Fourier coefficient of f is zero ($f_0 + f_1 = 0$), and therefore division in the frequency domain will also fail. The water level method leads to division of the zero frequency term by ε rather than zero. Note that (4.29) with $z = \exp i\omega\Delta t$ leads to division of the zero-frequency term by $1 - \rho$ rather than zero, pointing to a connection between ρ and the water-level parameter. Recursion, using equation (2.13), is an effective way to deconvolve the first-difference filter, leading to the same formula as the infinite-length inverse filter.

4.3.3 *Wiener filters*

The fourth method of deconvolution, design of a causal filter by least squares, is treated as a problem in inverse theory in Section 10.2.

EXERCISES

4.1. A low-pass Butterworth filter has amplitude spectrum

$$|A(\omega)|^2 = \frac{1}{1 + (\omega/\omega_c)^n}$$

For $\omega_c = 2\pi$ find the amplitudes at $\omega = 4\pi, 6\pi, 8\pi$ for $n = 1, 2, 3$. How does the value of n affect the frequency content of the output above the cut-off?

4.2. Find the convolution of the sequence $\{f_k;\ k = 0, 1, \ldots, N - 1\}$ with (a) the spike $\{\delta_{m0};\ m = 0, 1, \ldots\}$ and (b) the spike $\{\delta_{mQ};\ m = 0, 1, \ldots\}$, with Q fixed.

4.3. (a) Is the sequence (2,−3,−3, 2) minimum phase? If not, find the minimum phase sequence with the same amplitude spectrum.

(b) Verify by direct calculation that the autocorrelations of the two sequences are the same. This is a general result: why?

(c) For each sequence calculate the partial sums

$$S_p = \sum_{k=0}^{p} a_k^2; \quad p = 0, 1, 2, 3$$

to show the minimum phase sequence has more of its energy concentrated at the front.

4.4. Show that a two-pass filter (in which convolution is done first with the filter coefficients, then with the same coefficients reversed in time) is the same as convolution with the autocorrelation of the one-pass filter.

4.5. **Computer exercise: Butterworth filters using** `pitsa`

(a) A convolution of any sequence with a spike returns the original sequence. To see the filter characteristics in the time domain, simply filter a spike, then take the FFT to see its amplitude and phase spectra. Using `pitsa`, the procedure is to generate a spike in the centre of the record with
`Utilities` → `Test signals` → `Spikes`.
You can use 1024 points with the spike at point 512. Filter the spike with a Butterworth low pass filter using
`Advanced Tools` → `Filter` → `Butterworth Low Pass`

Use 1 Hz, answer 'no' to the one- or two-pass question, and choose a few different of sections (between one and nine). Examine the resulting filters in the time domain. Compare the onset of the filtered signals.

(b) Take their FFT and examine how the steepness of the amplitude spectrum varies with the number of sections.

(c) Vary the cut-off frequency and examine the difference in both time and frequency domains.

(d) Use the forward–backward ('y') option for the two-section Butterworth low-pass filter. This performs two passes of the convolution and produces a zero-phase filter. Compare with the one-pass filters in both time and frequency domains spectra.

4.6. **Computer exercise: filtering a broadband seismogram**

(a) Read the data in the file `nov1` with
`Files/Traces` → `Retrieve Files` → `ASCII`
Skip the first 80 points (this is header information containing, e.g., earthquake location, time, characteristics of the seismometer, etc.); the sampling frequency is 50 Hz. You should now have a seismogram lasting 20 min. Identify the P, S and surface waves.

(b) High-pass filter with a Butterworth, four sections, cut-off frequency 2 Hz. Can you see the aftershock?

(c) If this is indeed an aftershock it will have the same P–S time as the main shock. Measure the P–S times by
`Routine Tools` → `Zoom` → `untapered`
Use `Double Cursor` and note the time appearing in the window as you move the cursor around. Zoom in if you want to read the time more accurately.

(d) Select the first P wave on the original broad band trace by zooming in. Filter the trace with a bandpass 0.5–1.5 Hz, four-section filter.

(e) Compare the onset of the resulting trace with the high-pass filtered trace (above 2 Hz), and try to read the onset time in each case. You should find a big difference (10 s).

(f) This is an example of dispersion caused by a wave travelling through a thin, fast layer (in this case the subducted Pacific plate). The 1 Hz arrival (seen on the bandpass trace) travels at more or less normal velocity. Think about whether you should have used one- or two-pass filters for any interpretation.

4.7. Calculate the cross correlation of the sequences (1, 2, 3) and (1, −2, 1), their autocorrelations, and their correlation coefficients.

4.8. (a) Verify the sequence (6, −5, 1) is minimum phase by factorizing the z-transform.

(b) Construct the inverse filter by expressing the reciprocal of the z-transform in partial fractions and expanding each term.

(c) Calculate the first 10 terms of the inverse filter (use a calculator), and find the ratio of the last term to the first. Has the series converged well enough for 10 terms to make a reasonable deconvolution filter?

4.9. **Computer exercise: correlation**

You will find cross correlation under the `Advanced tools` menu. Use the `unscaled` option.

(a) Generate a Ricker wavelet by
`Utilities` → `Test signals` → `Ricker`
You will find the wavelet fills the screen nicely at 0.3 Hz.

(b) Find its autocorrelation. Note that pitsa does not plot the negative terms but they appear (symmetrically) at the far end: note also the total time has doubled. You can move the zero lag of the autocorrelation to the middle of the trace by
`Utilities` → `Trace utilities` → `Single channel` → `Cyclic shift`

(c) Now generate a noisy Ricker wavelet and cross correlate with the noise free Ricker wavelet (they should both have the same dominant frequency). Observe the noise reduction. Try shifting the noisy Ricker wavelet in time and repeat the cross correlation.

4.10. Sketch the wavelet given by the time sequence $(0, 6, 0, -5, 0, 1, 0)$. The wavelet represents a seismic source time function and you wish to deconvolve it from a seismic reflection record.

(a) Find its autocorrelation. Explain briefly why cross correlating a seismogram with the input wavelet is useful in finding later arrivals.

(b) Write down the z-transform of the original wavelet.

(c) Factorise the z-transform, show it is not minimum delay, and construct the minimum delay wavelet with the same autocorrelation.

(d) Show how to construct the inverse filter using partial fractions (there is no need to do the expansion). Explain why the procedure only works for a minimum delay wavelet.

(e) In this case, what is the difference between convolving with the minimum delay wavelet rather than the original wavelet? How would this be expected to influence the deconvolved reflection record?

4.11. Write down the (infinite-length) inverse filter for the two-point moving average filter $\frac{1}{2}\{1, 1\}$. Discuss how best to construct an approximate finite-length filter to do the same job.

4.12. **Computer exercise: deconvolution**

You will find deconvolution under the `Advanced tools` menu. `pitsa` does deconvolution by division of Fourier transforms. It asks for the 'water level', which is the 'white noise' added to the spectrum to stabilise the process and essentially prevent you from dividing by zero. It asks for the water level in a rather obscure way, and part of this practical is to work out for yourself how to choose the water level appropriately.

(a) First generate something that looks like a seismogram with some reflections and noise by the following procedure.

(1) Generate a 1 Hz Ricker wavelet with
`Utilities → Test Signals → Ricker Wavelet.`
Default sampling rate and record length will do. Keep this trace.

(2) Generate a noisy series of spikes, using variable amplitudes, both positive and negative, on the same trace. Use spike amplitudes about 1 and noise level about 0.01 to start with.

(3) Convolve the Ricker wavelet with the noisy spikes. You can now imagine this series as a seismogram with a number of noisy arrivals.

(b) Now deconvolve the third trace with the original, noise-free Ricker wavelet. Choose the appropriate water level to give the best result. Add noise to the denominator and repeat.

(c) Finally, investigate a range of signal-to-noise ratios.

5

Processing two-dimensional data

5.1 The 2D Fourier transform

So far we have dealt with sequences that are digitised samples of a function of one variable, the time. The same techniques apply to functions of distance, when it is appropriate to refer to *wavenumber* rather than frequency. A gravity traverse, for example, yields data as a function of distance along the traverse. Time series methods can be generalised to sequences derived from digitised functions of two or more independent variables. For example, gravity data collected on an x, y-grid can be digitised as

$$a_{kl} = a(k\Delta x, l\Delta y), \tag{5.1}$$

as can seismic data from an equally spaced array:

$$a_{kl} = a\,(k\Delta t, l\Delta x)\,. \tag{5.2}$$

We can define a two-dimensional (2D) Discrete Fourier transform (2DDFT) by taking two DFTs. In the case of equation (5.2), the first one-dimensional (1D) transform is taken with respect to time and the second with respect to distance,

$$A_{nm} = \sum_{k=0}^{K-1} \sum_{l=0}^{L-1} a_{kl} e^{-2\pi i(kn/K + lm/L)}, \tag{5.3}$$

where n counts the frequency in units of $1/K\Delta t$ and m counts the wavenumber in units of $1/L\Delta x$. In the case of equation (5.1), for 2D spatial data, the 1D transforms are with respect to x and y, and n, m are wavenumbers for the x- and y-directions respectively. The 1D transforms may be taken in any order.

Since the 2D transform is made up of two successive 1D transforms, the inversion formula can be derived simply by inverting each 1D transform in turn,

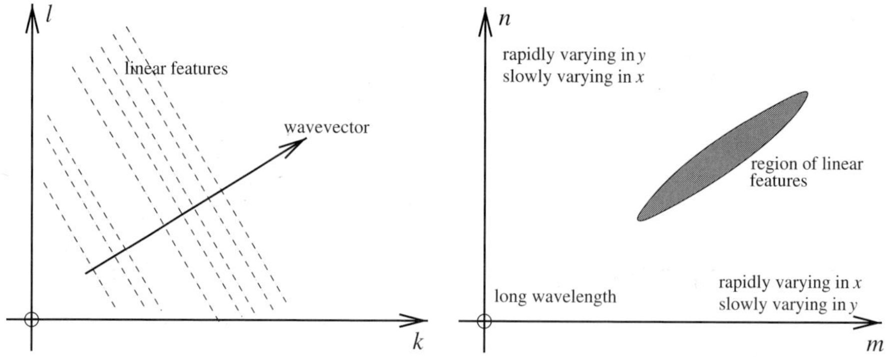

Fig. 5.1. Two-dimensional data can combine high and low wavenumbers, depending on the orientation of the structures. Linear patterns are characterised by constant wavenumber ratio m/n, even though they contain a range of wavenumbers. They map to a line in the transform domain at right angles to the lineation.

by using equation (2.18):

$$a_{kl} = \frac{1}{KL} \sum_{n=0}^{K-1} \sum_{m=0}^{L-1} A_{nm} e^{2\pi i(nk/K + ml/L)}. \tag{5.4}$$

Properties of the 2D transform may be derived from those of the 1D transform: there are two-dimensional shift, convolution, and Parseval theorems. Differentiation with respect to x is equivalent to multiplying the tranform by $2\pi i n/K\,\Delta x$ and differentiation with respect to y becomes multiplication by $2\pi i m/K\,\Delta y$.

The relationship between spatial anomalies and their transform is complicated by the presence of two wavenumbers. Data that vary rapidly in the x-direction may not vary much in y. Such data will look like lines parallel to the y-axis; they map into the lower right part of the transform domain because the x-wavenumber is large and the y-wavenumber is small (Figure 5.1). General lineations share the same ratio of wavenumber, $m/n =$ constant. The *wavevector* is defined as having components (n, m); for linear features the wavevector points perpendicular to the linearities, in the direction of a ray if the data are recording wavecrests (Figure 5.1).

Aliasing in two dimensions is slightly different from aliasing in one dimension. The Fourier transform is periodic in both directions:

$$A_{n+K,m+L} = A_{n+K,m} = A_{n,m+L} = A_{n,m}. \tag{5.5}$$

We can therefore 'tile' the frequency–wavenumber space as in Figure 5.1. For real data, half the transform coefficients within each tile are simply related to the other half

$$A_{K-n,L-m} = A^*_{n,m} \tag{5.6}$$

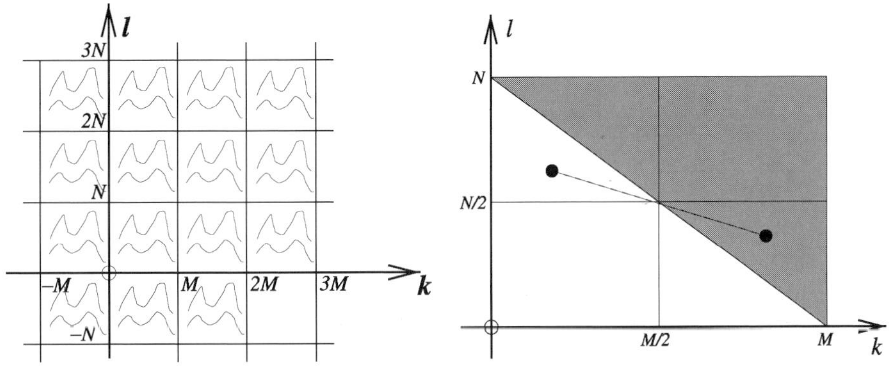

Fig. 5.2. Periodic repetition of the 2D discrete Fourier transform causes aliasing as in one dimension. Similar patterns in the basic rectangle set by the two sampling wavenumbers 'tile' the whole (k, l) domain (left figure). Folding occurs when the sum of wavenumbers is above the sum of the two Nyquist wavenumbers.

exactly as we should expect, having replaced KL real values with KL complex coefficients within each tile: half of the complex numbers are redundant. The symmetry of this form of aliasing is shown in Figure 5.2. Interestingly, it means that wavenumbers above the usual 1D Nyquist in the x-direction $(1/2\Delta x)$ are not aliased provided the wavenumber in the y-direction is sufficiently below its 1D Nyquist $(1/2\Delta y)$, and vice versa.

5.2 2D filtering

Filtering in 2D involves multiplication by a filter function in the frequency–wavenumber domain. Simple low-pass, bandpass, high-pass, and notch filters may be defined using the length of the wavevector $|\boldsymbol{n}| = \sqrt{m^2 + n^2}$. Thus a low-pass filter with cut-off N allows through all wavevectors inside the circle $\sqrt{m^2 + n^2} < N$; a high-pass filter allows through all wavevectors outside the circle; a bandpass filter allows through all wavevectors within an annulus; and a notch filter excludes all wavenumbers within the annulus. The same considerations of roll-off apply as in one dimension; for example we could construct a 2D Butterworth filter by using the function (4.1) with $|\boldsymbol{n}|$ replacing ω.

In 2D we are not restricted to circular filter functions, but any departure from circles means that one direction is filtered preferentially over another. There are sometimes good reasons for this: to clean up linear features, for example. It is sometimes desirable to look at the gradient of the data in a given direction, which is equivalent to multiplication by a wavenumber, or combination of the wavenumbers, in the wavenumber domain. Another example of anisotropic filtering is to separate waves travelling at different speeds (Section 5.3).

2D filters are important in processing potential data. The magnetic or gravitational potential, V, above the Earth's surface satisfies Laplace's equation:

$$\nabla^2 V = 0. \tag{5.7}$$

Take the Earth's surface to be at $z = 0$ in a cartesian coordinate system with z pointing upwards; equation (5.7) is solved by separation of variables by assuming a form

$$V(x, y, z) = X(x)Y(y)Z(z). \tag{5.8}$$

Substituting into (5.7) leads to

$$\frac{X''}{X} + \frac{Y''}{Y} + \frac{Z''}{Z} = 0. \tag{5.9}$$

Since each term is a function of a different variable, each must be constant. Choosing $X''/X = -m^2$ and $Y''/Y = -n^2$ leads to

$$X'' + m^2 X = 0 \tag{5.10}$$

$$Y'' + n^2 Y = 0 \tag{5.11}$$

$$Z'' - (n^2 + m^2)Z = 0 \tag{5.12}$$

The equations for X and Y give sinusoidal solutions while that for Z gives exponentials. We require V to decay with height because the sources are below ground, and we therefore eliminate the exponentially growing solution. Thus one solution of (5.7) is

$$X(x)Y(y)Z(z) = V_{mn} \mathrm{e}^{-p(m,n)z} \mathrm{e}^{\mathrm{i}(mx+ny)}, \tag{5.13}$$

where $p(m, n) = \sqrt{m^2 + n^2}$ is the positive decay rate with height. The general solution of (5.7) is a superposition of all such solutions:

$$V(x, y, z) = \sum_{m,n} V_{mn} \mathrm{e}^{-pz} \mathrm{e}^{\mathrm{i}(mx+ny)}. \tag{5.14}$$

The coefficients V_{mn} must be found from boundary conditions, which are provided by measurements. Putting $z = 0$ in (5.14) gives

$$V(x, y, 0) = \sum_{m,n} V_{mn} \mathrm{e}^{\mathrm{i}(mx+ny)}. \tag{5.15}$$

This has the form of a 2D Fourier series. If we make measurements at ground level, $z = 0$, we can find the V_{mn} from $V(x, y, 0)$ using the theory of Fourier series. Putting these back into (5.14) we can, in principle, find V at any height. This process is called *upward continuation*; it is useful, for example, in comparing data measured from an aircraft with ground data.

Clearly, equation (5.15) may be digitised to form a 2D Fourier inverse transform of the form (5.2). The V_{mn} then form the 2DDFT of spatial data measured at $z = 0$. Upward continuation to height H involves multiplying each Fourier coefficient by $\exp(-pH) = \exp(-|\boldsymbol{n}|^2 H)$. This is a form of low-pass filtering: the high wavenumbers are suppressed by the decreasing exponential. This is a very effective method for upward continuation. As the height H increases we move further from the sources of the magnetic or gravitational anomalies and small-scale detail is lost. The anomaly map at altitude therefore contains much less fine-scale detail than the corresponding map at ground level (Figure 5.3). Downward continuation is similar to upward continuation but suffers from instability. High wavenumbers are enhanced rather than suppressed. Since high wavenumbers are generally noisy, the noise is also enhanced.

5.3 Travelling waves

A seismic reflection line provides data that are equally spaced in both time and space. Technical details of acquisition and processing of seismic reflection data are discussed fully in texts on exploration geophysics such as Yilmaz (2001). The 2DDFT maps the time–distance plane into the frequency–wavenumber domain, often called the f–k plane. It is a decomposition of the wavefield into plane travelling waves because the exponential factor is a digitised form of the usual travelling waveform $\exp ik(x - ct)$, where k is the wavenumber and c is the phase speed. Plane waves travelling at the given speed c have constant k/l in the space-time domain and constant n/m in the frequency–wavenumber domain; amplitudes associated with waves travelling at a given speed therefore appear on straight lines passing through the origin, their slope depending on the phase speed (Figure 5.4).

The 2D transform separates the waves according to their speed as well as their frequency and wavelength, giving literally an additional dimension for processing. We can therefore design filters that pass only waves travelling at certain phase speeds. Aliasing in the f–k domain can also be thought of as aliasing the wave speed (see Figure 5.2 with axes replaced by f and k). A wave travelling so fast that one crest passes two geophones during one time sample interval will be aliased to a wave that travels more slowly. We can filter wave speeds by excluding wedge-shaped regions of f–k space. Figure 5.4 shows a wedge bounded by two lines of constant ν/k, or constant wave speed. These are sometimes called fan-filters for obvious reasons; they remove all waves travelling between two wave speeds.

Figure 5.5 gives a practical example of an f–k filter applied to seismic reflection data. Figure 5.5(a) shows the record of a noise test with spacing 10 m; the horizontal scale is distance in metres from the source and the vertical scale is time in seconds. Slow waves show up as steep lines, so the high energy seen at

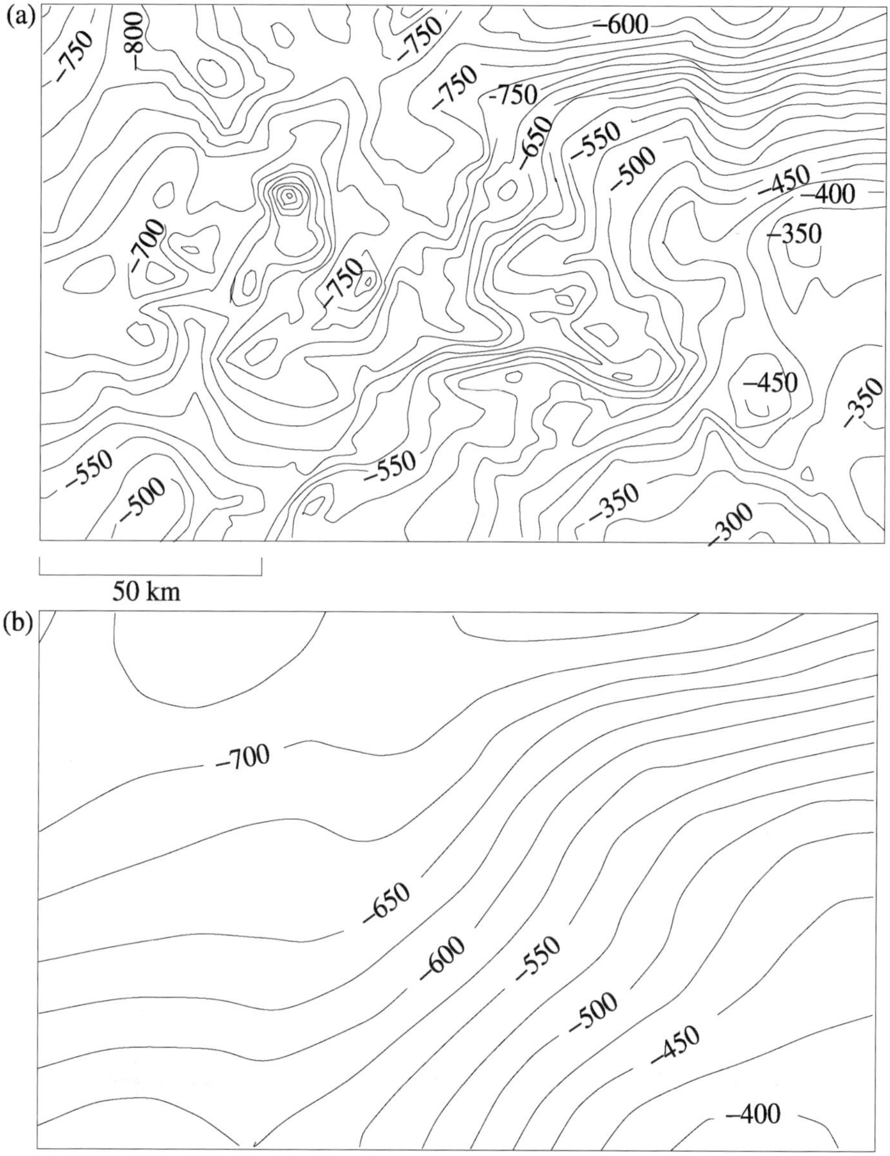

Fig. 5.3. (a) Bouger gravity anomalies at ground level and (b) the same anomalies upward continued to H =16 km. Redrawn, with permission, from Duncan and Garland (1977).

A is travelling more slowly than the energy seen at C or D. A is surface waves or 'ground roll', energy trapped in near-surface layers travelling at something like a Rayleigh wave speed, slower than either P or S body waves. B is backscattered energy; it arrives very late at short distance and must therefore have been reflected

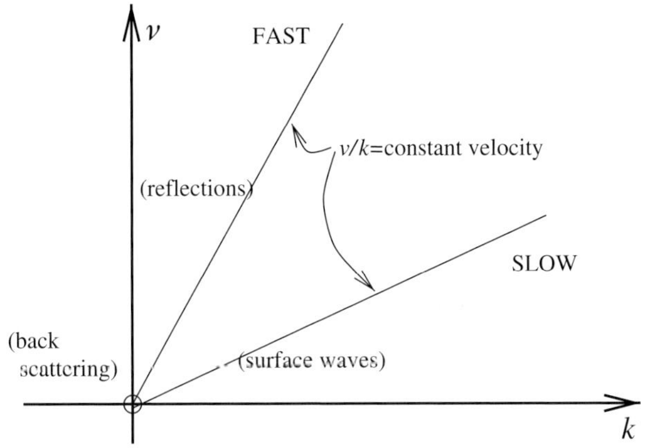

Fig. 5.4. The 2D discrete Fourier transform separates signals by phase speed. Fast wave energy (e.g. reflections) plots near the ν-axis while slow energy (e.g. surface waves) plots near the k-axis. It is therefore possible to separate the two even when they overlap in frequency and 1D filtering fails. Backwards-travelling waves plot with ν/k negative.

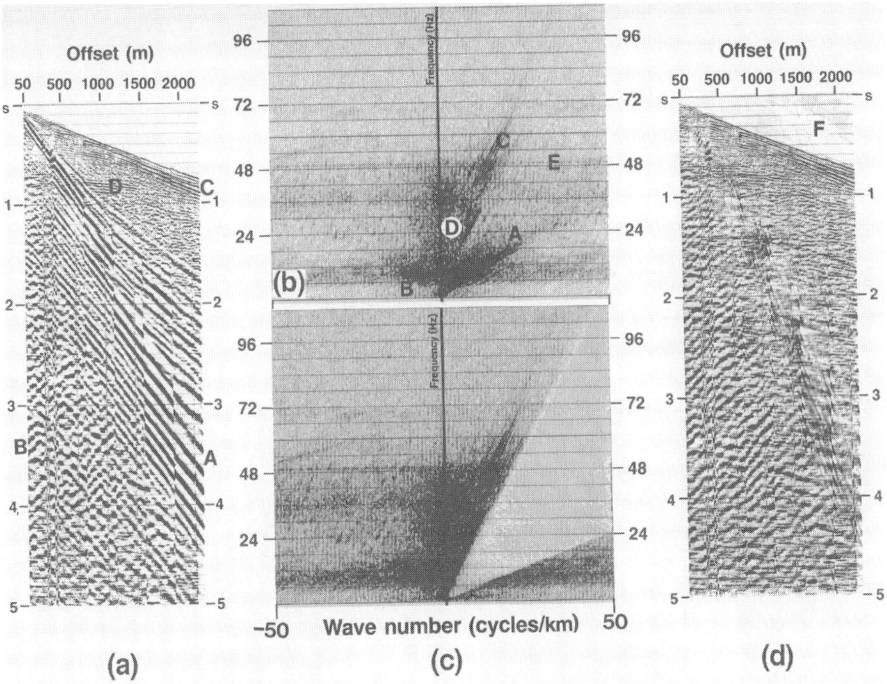

Fig. 5.5. Example of f–k filtering of a seismic reflection walk-away noise test. See text for details. Reproduced, with permission, from Yilmaz (2001).

by some anomalous structure. C is dispersive energy that also travels through near-surface layers. It travels fast because it follows the direct path from the source to the receivers at or near the P wave speed. D is the primary reflection, the main target of interest. It travels very quickly and is represented by nearly horizontal lines that do not intersect with the origin. The high speed is caused by near-vertical incidence of the reflected rays; this makes the wavefront sweep across the ground very quickly. The lines do not pass through the origin because of the time taken for the wave to travel down to the reflector and back: at normal incidence the wavefront travels across the ground at infinite speed and so appears as a horizontal line at time t, the two-way travel time from surface to reflector.

Figure 5.5(b) shows the 2DDFT of (a), with f along the vertical axis and k along the horizontal axis. The ground roll is at A, with low values of f/k because of the low wave speed. B appears at negative values of k because it is travelling backwards. The dispersed waves C are high wavenumber and high frequency; they travel faster than A and are at a higher value of f/k. Faster still, and with lower f and k, are the primary reflections. Figure 5.5(c) shows the application of a 'dip' or fan filter designed to remove the ground roll A. Figure 5.5(d) shows the inverse transform of (c) with the ground roll removed. The fan filter has done nothing to remove the dispersive waves C because these travel at speeds outside the rejection wedge of the filter.

There is no reason to restrict the filter shape to a simple wedge (or circle, as in the example of low-pass filtering). In this example we could design a filter to remove the dispersive waves also, leaving the primary reflections, by excluding a polygon shape containing C and E. Modern processing software allows the operator to draw a filter shape with the mouse before inverting the transform.

2D filtering suffers the same restrictions as 1D filtering. The edges of the wedge must not be too sharp but must be tapered to avoid ringing effects and leakage, just as for the Butterworth filter in 1D. The fan width must not be too narrow. A narrow width has the same effect as a very narrow bandpass filter, which produces a near-monochromatic wave that can be recovered accurately only with a very long time series.

EXERCISES

5.1. Take the 2D discrete Fourier transform of the following array.

	$k=0$	$k=1$	$k=2$
$l=0$	0	0	1
$l=1$	0	1	0
$l=2$	1	0	0

(a) Sketch the pattern of aliased frequencies.

(b) Compute all nine values of the transform and point out the aliased values. (Beware: many values are coincidentally the same but are not aliased.)
(c) Find the amplitude spectrum.

5.2. A wave travelling at speed c with wavelength $2\pi/k$ has the form

$$A \cos 2\pi k(x - ct).$$

By writing the cosine in terms of complex exponentials, find the two non-zero coefficients of the complex 2D Fourier transform of the wave.

5.3. You collect data from a linear array of geophones with spacing 1 km and sampling interval 25 ms.

(a) Calculate the Nyquist frequency and Nyquist wavelength.
(b) A wave sweeps across the array with plane wavefront and apparent velocity 5 km s^{-1}. Show with a diagram where this wave energy would appear in the 2D Fourier transform of the data.
(c) Show also where you would expect the aliased wave energy to appear.
(d) You wish to design an f–k filter to remove microseismic noise but retain body waves. The microseisms travel slower than 3 km s^{-1} with periods in the range 2–8 s. The body waves travel faster than 3 km s^{-1} and have frequencies in the range 0.25–5 Hz. Sketch the regions in the 2D transform domain where you expect to find the body waves and microseisms, and outline the characteristics of a filter to separate the two.
(e) Discuss the effectiveness of this filter when the microseisms are larger in amplitude than the body waves, bearing in mind the need for roll-off at the cut-offs.

5.4. You carry out a magnetic survey of a square field with side 100 m in search of signals from buried Roman buildings. The ambient noise has a dominant wavelength of about 0.3 m.

Give your choice of sampling interval (in metres) and outline how you would design a 2D low-pass Butterworth filter to enhance the signal from the buildings and streets. Give the mathematical expression for the amplitude response of the filter in the wavenumber domain.

Part II

Inversion

6

Linear parameter estimation

6.1 The linear problem

We parametrise the model with a set of P parameters assembled into a *model vector*

$$\boldsymbol{m}^{\mathrm{T}} = (m_1, m_2, \ldots, m_P). \tag{6.1}$$

The parametrisation depends on the specific problem: examples will follow. The data are put into a *data vector* of length D:

$$\boldsymbol{d}^{\mathrm{T}} = (d_1, d_2, \ldots, d_D) \tag{6.2}$$

with corresponding errors, also placed into a vector of length D:

$$\boldsymbol{e}^{\mathrm{T}} = (e_1, e_2, \ldots, e_D). \tag{6.3}$$

Note that all scientific measurements have an error attached to them. Without the error they are not scientific measurements.

When the forward problem is linear the relationship between data and model vectors is linear and may therefore be written in the matrix form

$$\boldsymbol{d} = \mathsf{A}\boldsymbol{m} + \boldsymbol{e}, \tag{6.4}$$

where A is a $D \times P$ matrix of coefficients that are independent of both data and model. The elements of A must be found from the formulation of the forward problem: they depend on the physics of the problem and geometry of the experiment, the type of measurements rather than their accuracy. Equations (6.4) are called the *equations of condition*; they may be written out in full as

$$d_i = \sum_{j=1}^{P} A_{ij} m_j + e_i. \tag{6.5}$$

Table 6.1. *Many important geophysical inverse problems are linear. The coefficients of matrix* A *must be derived from the forward problem, which usually involves basic laws of physics*

Problem	Data	Model	A
Gravity interpretation	g	density	Newton's law of gravity
Geomagnetic main field modelling	field components X, Y, Z	geomagnetic coefficients	potential theory
Downward continuation	$V(H)$	$V(0)$	potential theory
Deconvolution	desired output	filter coefficients	convolution
Seismometer response	seismometer output	impulse response	convolution
Finding core motions	secular variation	fluid velocity	Maxwell's equations
Heat flow	temperature	heat source	heat conduction

In the language of inverse theory, the ith row of A is called the *data kernel*; it describes how the ith datum depends on the model. Some real examples of geophysical linear inverse problems are given in Table 6.1.

If $D = P$ the number of equations is equal to the number of unknowns, A is square, and provided $\det \mathsf{A} \neq 0$ the model may be found by the usual methods for simple simultaneous equations or using the inverse of the matrix. Such a problem is said to be *equi-determined*. If $D > P$ and there are more independent equations than unknowns the problem is said to be *over-determined*. If there are no errors in the data the equations must be compatible for a solution to exist. For example, the equations $x = 2$ and $2x = 4$ are compatible; $x = 2$ and $x = 3$ are not. In the presence of errors a range of solutions are possible. In Section 6.2 we minimise the errors to find an optimal solution. When $D < P$ there are fewer linearly independent equations than unknowns the problem is said to be *under-determined*. It is not possible to solve for all the model parameters uniquely. Again a range of solutions are possible. Evaluating them requires inversion proper, as described in Chapter 7.

The basic ideas are illustrated by a very simple inverse problem: given measurements of the Earth's mass (M) and moment of inertia (I), what can we discover about the density? There are two data, $D = 2$:

$$M = 5.974 \times 10^{24}\ \mathrm{kg} \tag{6.6}$$

$$I/a^2 = 1.975 \times 10^{24}\ \mathrm{kg}, \tag{6.7}$$

where a is the Earth's radius. The errors are ± 3 in the last decimal place in each case.

Next we need a model for the density. We shall make a number of restrictive assumptions, none of which is strictly true and some of which are clearly unrealistic,

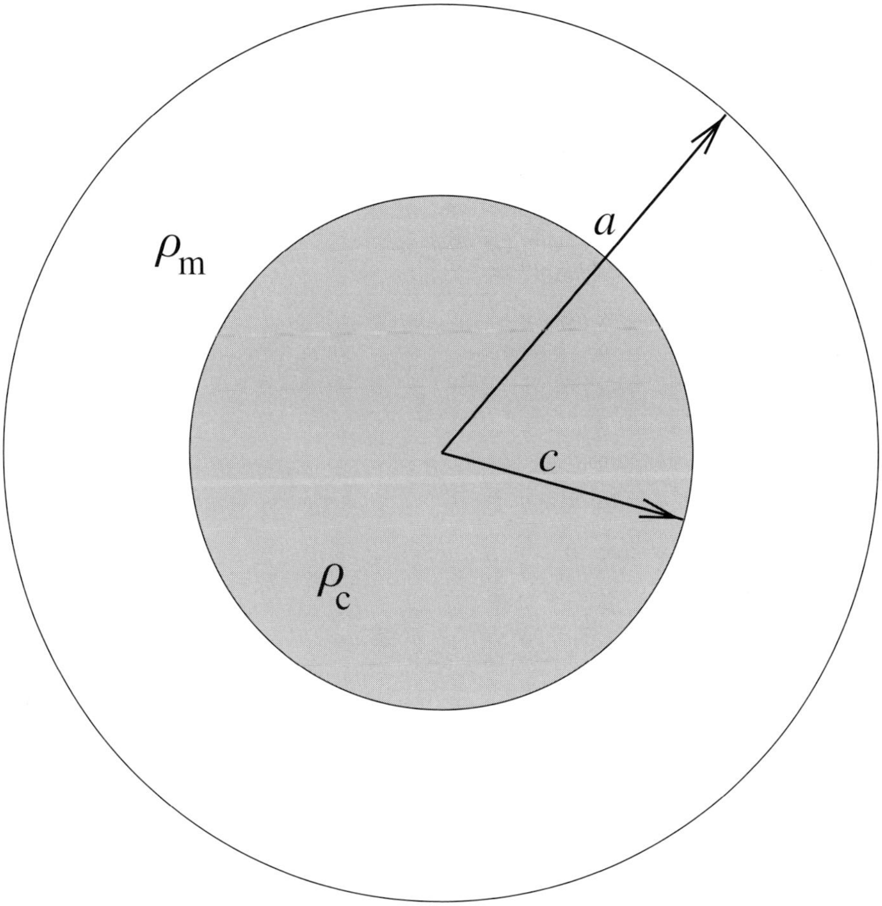

Fig. 6.1. The model used to invert mass and moment of the Earth for density. The Earth is assumed to be spherically symmetric; the core and mantle are assumed to have known radii and unknown uniform densities.

but they serve to illustrate the methods. We assume that the Earth is spherically symmetric, so that density is a function of radius alone (Figure 6.1), that the radius is known exactly (6371 km) from other measurements, and that the Earth has two distinct regions, the mantle and core, whose radius c is also known exactly (3485 km). For the present illustrations we also make the unrealistic assumption that the densities of mantle and core are constants, independent of radius, and for some models the two densities are equal, giving effectively a uniform Earth. There are six different problems, given in Table 6.2. Problems (1), (2), and (4) are equidimensional and will be solved here, problem (3) is over-determined and will be solved in the next section, and problems (5) and (6) are under-determined and will be solved in Chapter 7.

Table 6.2. *Different problems to find the Earth's density from its mass and moment of inertia*

(1) M assuming uniform Earth	$D=1$	$M=1$
(2) I/a^2 assuming uniform Earth	$D=1$	$M=1$
(3) M and I/a^2 assuming uniform Earth	$D=2$	$M=1$
(4) M and I/a^2 assuming two-layer Earth	$D=2$	$M=2$
(5) M assuming two-layer Earth	$D=1$	$M=2$
(6) I assuming two-layer Earth	$D=1$	$M=2$

The forward problem is given by the simple integral relationships between the mass and moment of inertia and the density as a function of radius:

$$M = 4\pi \int_0^a \rho(r) r^2 \mathrm{d}r \tag{6.8}$$

$$\frac{I}{a^2} = \frac{8\pi}{3a^2} \int_0^a \rho(r) r^4 \mathrm{d}r. \tag{6.9}$$

For an Earth of constant density $\bar{\rho}$, (6.8) gives

$$M = \frac{4\pi}{3} \bar{\rho} a^3 = 5.974 \times 10^{24}\ \mathrm{kg}, \tag{6.10}$$

which is easily solved to give

$$\bar{\rho} = \frac{3}{4\pi a^3} M = 5.515 \times 10^3\ \mathrm{kg}. \tag{6.11}$$

The error on M maps to $\bar{\rho}$ by the same formula to give

$$\bar{\rho} = (5.515 \pm 0.003) \times 10^3\ \mathrm{kg\,m^{-3}}. \tag{6.12}$$

This is the solution to the equi-determined problem (1).

Problem (2) needs the formula for the moment of inertia of a uniform sphere, derived from (6.9),

$$\frac{I}{a^2} = \frac{8\pi}{15} \bar{\rho} a^3 = 0.4M, \tag{6.13}$$

with solution

$$\bar{\rho} = \frac{15 I}{8\pi a^5} = 4.558 \times 10^3\ \mathrm{kg}. \tag{6.14}$$

The error in I maps to $\bar{\rho}$ by the same formula to give

$$\bar{\rho} + (4.558 \pm 0.007) \times 10^3\ \mathrm{kg\,m^{-3}}. \tag{6.15}$$

Clearly something is wrong here – the two independent estimates of density differ by about 10^3 kg m^{-3}, a hundred times larger than would be expected from the error estimates!

Obviously the data are incompatible with the model. The model is too simple. Density varies as a function of radius throughout the mantle and core. This illustrates the difficulty with most geophysical problems: data error is often unimportant in relation to errors arising from uncertainties about the assumptions behind the model.

We now make the model more complicated by allowing different densities for core and mantle. This is problem (4), also equi-determined. Equations (6.8) and (6.9) give

$$M = \frac{4\pi}{3} c^3 \rho_c + \frac{4\pi}{3}(a^3 - c^3)\rho_m \tag{6.16}$$

$$\frac{I}{a^2} = \frac{8\pi}{15}\frac{c^5}{a^2}\rho_c + \frac{8\pi}{15}\left(a^3 - \frac{c^5}{a^2}\right)\rho_m. \tag{6.17}$$

These equations are in the form of (6.4). The data vector is $\boldsymbol{d}^T = (M, I/a^2)$; the model vector is $\boldsymbol{m}^T = (\rho_c, \rho_m)$, and the equations of condition matrix is

$$\mathsf{A} = \frac{4\pi}{3}\begin{pmatrix} c^3 & (a^3 - c^3) \\ \frac{2}{5}\frac{c^5}{a^2} & \frac{2}{5}\left(a^3 - \frac{c^5}{a^2}\right) \end{pmatrix}. \tag{6.18}$$

Note that A depends only on the two radii, not on the densities nor the data. Solving these two simple simultaneous equations is straightforward and gives

$$\rho_c = 12.492 \times 10^3 \text{ kg m}^{-3} \tag{6.19}$$

$$\rho_m = 4.150 \times 10^3 \text{ kg m}^{-3}. \tag{6.20}$$

The core is denser than the mantle, as it must be for gravitational stability. There is no incompatibility and the model predicts the data exactly. There is no redundancy in the data and we can do no further tests to determine whether this is a realistic model for the Earth's density. The errors are estimated in Section 6.4.

6.2 Least-squares solution of over-determined problems

In the over-determined case there are more equations than unknowns. The least-squares solution minimises the sum of squares of the errors. From (6.5):

$$e_i = d_i - \sum_{j=1}^{P} A_{ij} m_j \tag{6.21}$$

and the sum of squares is

$$E^2 = \sum_{i=1}^{D} e_i^2 = \sum_{i=1}^{D} \left(d_i - \sum_{j=1}^{P} A_{ij} m_j \right)^2 . \tag{6.22}$$

We choose the unknowns $\{m_k\}$ to minimise E by setting $\partial E^2/\partial m_k = 0$

$$2 \sum_{i=1}^{D} \left[d_i - \sum_{j=1}^{P} A_{ij} m_j \right] (-A_{ik}) = 0, \tag{6.23}$$

which may be rearranged as

$$\sum_{j=1}^{P} \sum_{i=1}^{D} A_{ik} A_{ij} m_j = \sum_{i=1}^{D} A_{ik} d_i . \tag{6.24}$$

In matrix notation this equation may be written

$$\mathsf{A}^{\mathrm{T}} \mathsf{A} \mathbf{m} = \mathsf{A}^{\mathrm{T}} \mathbf{d}. \tag{6.25}$$

These are called the *normal equations.*

If $D > P$ then $\mathsf{A}^{\mathrm{T}}\mathsf{A}$ is symmetric and positive semi-definite (all eigenvalues are positive or zero). If there are no zero eigenvalues then $\mathsf{A}^{\mathrm{T}}\mathsf{A}$ has an inverse and (6.25) may be solved for $\boldsymbol{m}$. Writing the least-squares solution of (6.4) as

$$\boldsymbol{m} = (\mathsf{A}^{\mathrm{T}}\mathsf{A})^{-1}\mathsf{A}^{\mathrm{T}}\boldsymbol{d}. \tag{6.26}$$

We can regard the $P \times D$ rectangular matrix $(\mathsf{A}^{\mathrm{T}}\mathsf{A})^{-1}\mathsf{A}^{\mathrm{T}}$ as a sort of inverse of the $D \times P$ rectangular matrix A: it is the *least-squares inverse* of A (Appendix 4). When $D = P$ the matrix is square and the least-squares solution reduces to the usual form $\boldsymbol{m} = \mathsf{A}^{-1}\boldsymbol{d}$.

Now apply the least-squares method to problem (3) in Table 6.2, for a uniform Earth with two data. The two forward modelling equations are

$$M = \frac{4\pi}{3} a^3 \bar{\rho} + e_M \tag{6.27}$$

$$\frac{I}{a^2} = \frac{8\pi}{15} a^3 \bar{\rho} + e_I, \tag{6.28}$$

where e_M, e_I are the errors on M and I/a^2 respectively. We wish to find $\bar{\rho}$ such that $E^2 = e_M^2 + e_I^2$ is a minimum:

$$E^2 = \left(M - \frac{4\pi}{3} a^3 \bar{\rho} \right)^2 + \left(\frac{I}{a^2} - \frac{8\pi}{15} a^3 \bar{\rho} \right)^2 . \tag{6.29}$$

Differentiating with respect to the unknown $\bar{\rho}$ gives

$$\frac{\partial E^2}{\partial \bar{\rho}} = -2\frac{4\pi}{3}a^3\left(M - \frac{4\pi}{3}a^3\bar{\rho}\right) - 2\frac{8\pi}{15}a^3\left(\frac{I}{a^2} - \frac{8\pi}{15}a^3\bar{\rho}\right) = 0, \tag{6.30}$$

which may be simplified to give

$$\bar{\rho} = \frac{(M + 2I/5a^2)}{\left(\frac{4\pi}{3}a^3\right)\frac{29}{25}} = 5.383 \times 10^3\ \mathrm{kg\,m^{-3}}. \tag{6.31}$$

The 2×1 equations-of-condition matrix is

$$\mathsf{A} = \left(\frac{4\pi}{3}a^3, \frac{8\pi}{15}a^3\right). \tag{6.32}$$

The normal-equations matrix in (6.25) becomes the 1×1 matrix (or scalar):

$$\mathsf{A}^{\mathrm{T}}\mathsf{A} = \frac{29}{25}\left(\frac{4\pi}{3}a^3\right)^2. \tag{6.33}$$

The right-hand sides of the normal equations matrix become

$$\mathsf{A}^{\mathrm{T}}\boldsymbol{d} = \left(\frac{4\pi a^3}{3}\right)\left(1, \frac{2}{5}\right)\begin{pmatrix} M \\ I/a^2 \end{pmatrix} = \left(\frac{4\pi a^3}{3}\right)\left(M + \frac{2I}{5a^2}\right). \tag{6.34}$$

Since $\mathsf{A}^{\mathrm{T}}\mathsf{A}$ is a scalar, matrix inversion is trivially division by a scalar, leaving

$$\boldsymbol{m} = \frac{(M + 2I/5a^2)}{\left(\frac{4\pi}{3}a^3\right), \frac{29}{25}} \tag{6.35}$$

in agreement with (6.31).

So far we have been using I/a^2 as data. What happens if we use I instead? For the equi-determined problems this obviously makes no difference; the only effect is to multiply one of the equations by a^2 and the solution is unchanged. For the least-squares solution there is a difference because a different quantity is minimised. Multiplying equation (6.28) by a^2 changes (6.29) to

$$E^2 = \left(M - \frac{4\pi}{3}a^3\bar{\rho}\right)^2 + \left(I - \frac{8\pi}{15}a^5\bar{\rho}\right)^2. \tag{6.36}$$

Working through to the solution with this new definition of E gives

$$\bar{\rho} = \frac{(M + 2Ia^2/5)}{\frac{4\pi}{3}a^3\left(1 + \frac{4}{25}a^4\right)} = 4.558 \times 10^3 \text{ kg m}^{-3}. \tag{6.37}$$

Verification of this formula is left as an exercise at the end of this chapter.

The solution (6.37) is different from (6.31). It is the same, to the accuracy presented, as (6.14). Comparing (6.29) and (6.36) explains why: the contribution of the second term in (6.36) has been multiplied by a^4 in metres, or 1.895×10^{22}! The first term in (6.36), the contribution from M, is negligible in comparison to the second contribution, that from I. M contributes only to the final solution (6.37) after about the 20th significant figure! $\bar{\rho}$ is determined only by I, as was (6.14).

Scaling the equations of condition makes a difference to the solution of the normal equations, and the scaling must therefore be chosen carefully. In this example we mixed different types of data of different dimensions. Dividing I by a^2 gives a quantity with the dimensions of mass. This is a good general guide: before performing an inversion reduce the data to the same dimension using some appropriate scales of length, mass, time, etc. In the next section we shall see the value of dividing all data by their errors so that each measurement contributes equally to the solution of the normal equations. This removes any problems with inhomogeneous datasets.

In the formalism, converting the data from I to I/a^2 involves dividing both sides of equation (6.28) by a^2. The elements of the equations of condition matrix in row 2 are divided by a^2. The square of the element in each row contributes to all the elements of the normal equations matrix $\mathsf{A}^T\mathsf{A}$. Reducing the whole row reduces its contribution to the normal equations and reduces the importance of that particular datum to the solution.

6.3 Weighting the data

Estimating the error on a measurement is a vital part of the experimental process. For inversion, the 'error' is any contribution to the data that does not satisfy the forward modelling equations. Many different sources can contribute to the error. The most obvious, and easiest to deal with, is instrument error. As an example, in 1623 Jacques l'Hermite travelled from Sierra Leone via Cape Horn observing declination with two compasses, one calibrated in Amsterdam and the other in Rotterdam. The two instruments drifted by $1^\circ 12'$ on average. This value was taken by Hutcheson and Gubbins (1990) to be representative of the error in declination at sea during the seventeenth century.

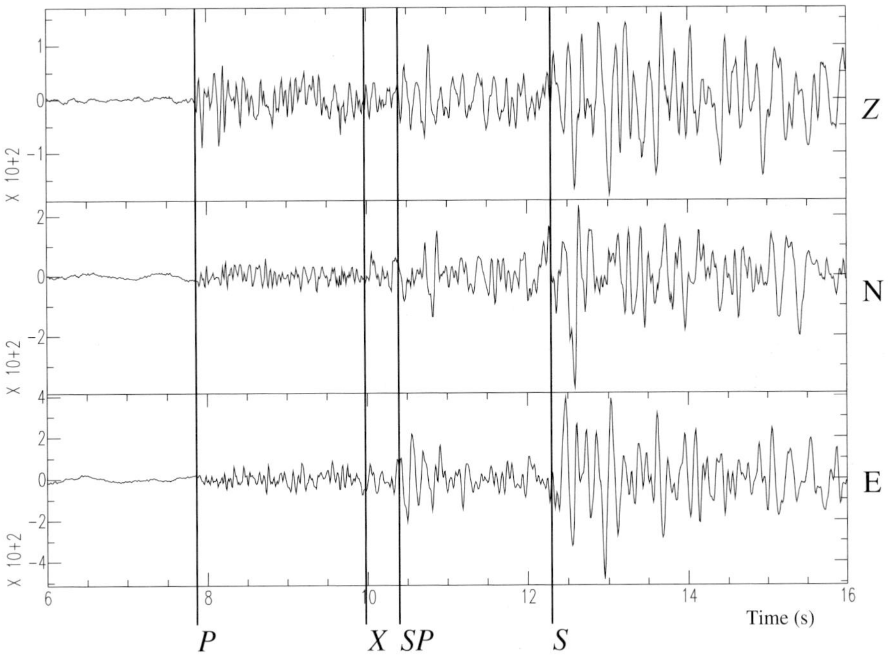

Fig. 6.2. Seismogram recorded in North Island, New Zealand, from a local earthquake beneath Cook Strait. Errors on arrival time picks depend on the type of arrival and signal strength in relation to the background noise. Here the *P* wave will be most accurately read on the vertical (*Z*) component, the *S* wave on the East component, and the intermediate phases (probably conversions from *S* to *P* at a refracting interface beneath the station) on the vertical. Error estimates on the picks should reflect this: *S* waves are usually given a larger error than *P* simply because of their longer period.

Geophysics suffers particularly from errors that are not directly related to the measurement process. The accuracy of the onset time of a seismic phase may depend on the pulse shape or the presence of other arrivals in the same time window, as well as the ambient noise level, which itself can vary dramatically with time (Figure 6.2). In this case we might estimate the errors at 0.02 s for *P*, 0.05 s for *X*, and 0.1s for *SP* and *S*. In practice one would develop a rough idea of the relative errors for these phases over a number of seismograms and apply the same error to each.

A measurement of the geomagnetic main field will be affected by the magnetisation of neighbouring crustal rocks. Near a seamount the contribution from highly magnetic lavas can be several degrees; an average figure for typical rocks is nearer one quarter of a degree, less than Jacques l'Hermite's instrumental error, but not by much. Crustal magnetisation limits the additional value of accurate, modern, marine data in determining the main field.

Then there are blunders: transcription errors such as a wrong sign in the latitude of a marine measurement, misreading the seismograph's clock by a whole minute, or wiring the seismometer up with the wrong polarity, thus changing the sign of each component. These blunders can have a disastrous effect on the whole inversion if they are not removed from the dataset altogether.

Errors are correctly treated as random variables. Their actual values remain indeterminate; we can only find certain statistical properties of the random variables such as the expected value, variances, and covariances of pairs of random variables (equations (2.42) and (2.43)). For *independent* random variables the covariance is zero. The *correlation coefficient* is a normalised form of the covariance

$$c(X, Y) = \frac{\mathcal{C}(X, Y)}{\sqrt{\mathcal{V}(X)\mathcal{V}(Y)}}, \tag{6.38}$$

which is equal to unity when the two variables are equal and therefore perfectly correlated, zero when they are uncorrelated.

The expectation operator is precisely defined in the theory of statistics but for our purposes we shall either (a) estimate it by taking averages of measurements, in which case the expected value is also the mean, or (b) impose it *a priori* from theory. In Chapter 2 we used the average over time. This kind of error estimation often reduces to little more than guesswork, but it has to be done.

The square root of the variance is called the *standard deviation* and is a measure of the 'scatter' of the data about the mean; for us it will be the error. If the variance is to be estimated by averaging a number of data $\{X_i; i = 1, 2, \ldots, N\}$ the mean is $\bar{X} = N^{-1} \sum X_i$ and the standard deviation is

$$\sigma = \sqrt{\frac{1}{N-1} \sum_{i=1}^{N} (X_i - \bar{X})^2}. \tag{6.39}$$

The divisor $(N - 1)$ is used rather than N because we have used up one 'degree of freedom' in estimating the mean. If the mean is known *a priori* then N should be used instead. The standard deviation of the mean is reduced by (approximately) the square root of the number of data. Table 6.3 gives examples of computing the mean and standard deviation of a set of measurements.

Table 6.4 gives a calculation of the covariance matrix. The data are artificial: the numbers have been calculated from the formula $Y = 2X + 1$ using random numbers for X to produce perfectly correlated data. The covariance matrix is

$$C_{XY} = \begin{pmatrix} 1.329 & 2.902 \\ 2.902 & 6.358 \end{pmatrix} \tag{6.40}$$

Table 6.3. *Computing the mean and variance from a set of six measurements by averaging*

X	$X - \bar{X}$
0.731	−0.030
0.765	0.004
0.752	−0.009
0.797	0.036
0.772	0.011
0.746	−0.015
$\bar{X} = 0.761$	$\mathcal{V}(X) = 0.000528$

Table 6.4. *Estimating the covariance of data by averaging. These numbers were made up from the formula $Y = 2X + 1$ using random numbers for X. They therefore correlate perfectly*

X	Y	$X - \bar{X}$	$Y - \bar{Y}$	$(X - \bar{X})(Y - \bar{Y})$
0.0881	0.7545	−1.3609	−2.9869	4.065
1.0181	2.9059	−0.4309	−0.8355	0.360
1.9308	4.6156	0.4818	0.8742	0.421
2.7588	6.6895	1.3098	2.9481	3.861
$\bar{X} = 1.4490$	$\bar{Y} = 3.7414$	$\mathcal{V} = 1.329$	$\mathcal{V} = 6.358$	$\mathcal{C} = 2.902$

and the correlation coefficient is

$$\frac{2.902}{\sqrt{1.329 \times 6.358}} = 1.00.$$

Another geophysical example of error estimation is shown in Figure 6.3. Permanent magnetic observatories like this one report annual means of the geomagnetic vector. A sudden change in the second time derivative was noted worldwide in 1969–70; it was called a geomagnetic jerk (Ducruix *et al.*, 1980). It shows up well when plotting first differences of the annual means, an estimate of the local secular variation. The accuracy of the individual differences can be estimated from the scatter about a smooth time variation before and after the jerk. The standard deviation of the residual after the straight-line prediction has been subtracted from the data to give the estimate of the accuracy of each annual mean. A typical observatory annual mean error is $5\,\mathrm{nT\,yr^{-1}}$. The contribution from magnetised crust is constant in time and will be removed by the differencing.

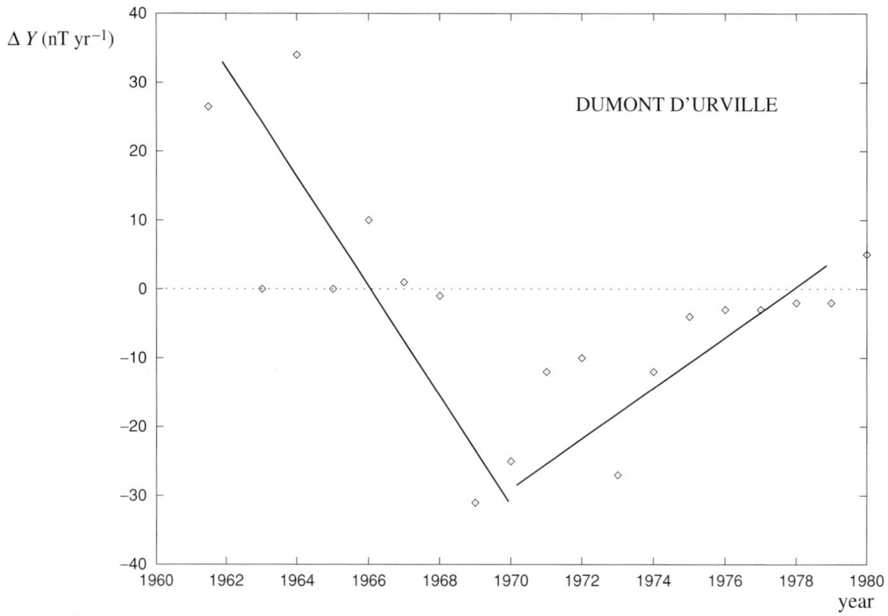

Fig. 6.3. First differences of annual means of geomagnetic East component measured at the Antarctic observatory Dumont D'Urville from 1960 to 1980. The steep change in slope is believed to be caused by sudden changes in the Earth's core, and is called a geomagnetic jerk. Measurement error is estimated from the scatter about the pair of straight lines.

The mean of the sum of two random variables is the sum of the means: $E[X + Y] = E[X] + E[Y]$. The variance of the sum depends on the covariance:

$$\begin{aligned}\mathcal{V}[X+Y] &= E[(X+Y-E[(X+Y)])^2] \\ &= E[((X-E[X]) + (Y-E[Y]))^2] \\ &= E[(X-E[X])^2 + 2(X-E[X])(Y-E[Y]) + (Y-E[Y])^2] \\ &= E[(X-E[X])^2] + 2E[(X-E[X])(Y-E[Y])] \\ &\quad + E[(Y-E[Y])^2]. \end{aligned} \tag{6.41}$$

So

$$\mathcal{V}[X+Y] = \mathcal{V}[X] + 2\mathcal{C}[X,Y] + \mathcal{V}[Y]. \tag{6.42}$$

For independent random variables the variances add because the covariance is zero:

$$\mathcal{V}[(X+Y)] = \mathcal{V}[X] + \mathcal{V}[Y]. \tag{6.43}$$

These ideas apply to an arbitrary number of data. Let $\boldsymbol{X}$ be an M-length vector whose elements are the random variables $X_1, X_2, \ldots, X_M$, and define a second vector $\boldsymbol{x}$ with zero mean

$$\boldsymbol{x} = \boldsymbol{X} - E[\boldsymbol{X}]. \tag{6.44}$$

The covariance matrix becomes, in matrix notation,

$$\mathcal{C}(\boldsymbol{x}) = E[\boldsymbol{x}\boldsymbol{x}^{\mathrm{T}}]. \tag{6.45}$$

Another view of variance is through the *likelihood function*. Consider a single random variable x. The likelihood function $L(x)$ depends on the probability of the variable taking on the value x. It is a maximum at x_L, when $\mathrm{d}L/\mathrm{d}x = 0$. Expanding in a Taylor series about $x = x_L$ gives

$$L(x) = L(x_L) + \frac{1}{2} \left.\frac{\mathrm{d}^2 L}{\mathrm{d}x^2}\right|_{x_L} (x - x_L)^2 + \cdots. \tag{6.46}$$

The variance of x may then be defined in terms of the value of $x - x_L$ when the likelihood function reaches some critical value. Provided higher terms in (6.46) are negligible, the variance will be proportional to the $\mathrm{d}^2L/\mathrm{d}x^2$. If x has a Gaussian distribution, with probability $p(x)$ proportional to $\exp[-(x - x_L)^2]/\sigma^2$, and the likelihood function is defined as $\log p$, then (6.46) has no higher terms.

For a vector $\boldsymbol{x}$ of many random variables the likelihood function may be expanded in a similar way

$$L(\boldsymbol{x}) = L(\boldsymbol{x}_L) + \frac{1}{2} \sum_{ij} \left.\frac{\partial^2 L}{\partial x_i x_j}\right|_{\boldsymbol{x}_L} (\boldsymbol{x} - \boldsymbol{x}_L)_i (\boldsymbol{x} - \boldsymbol{x}_L)_j + \cdots. \tag{6.47}$$

The covariance matrix can again be related to the second derivative

$$(C)^{-1}_{ij} = \frac{\partial^2 L}{\partial x_i x_j}. \tag{6.48}$$

Further details are in Tarantola (1987).

It is unusual to be able to estimate covariances directly for real data; most measurements are assumed to be independent. Correlations enter when the data are transformed in some way. For example, it is common to want to rotate the horizontal components of a seismogram from the conventional North–South and East–West into radial (away from the source) and transverse. This involves multiplying by a rotation matrix; the new components are linear combinations of the originals.

This always introduces off-diagonal elements into the covariance matrix and correlations between the new components.

Consider a new dataset formed by a general linear transformation of $\boldsymbol{x}$,

$$\boldsymbol{y} = \mathsf{G}\boldsymbol{x}, \tag{6.49}$$

where G is any matrix with compatible dimensions (it need not be square) such that $\boldsymbol{y}$ has zero mean when $\boldsymbol{x}$ has. The covariance matrix of $\boldsymbol{y}$ is found by substituting for $\boldsymbol{y}$ from (6.49) into (6.45) to give

$$\begin{aligned} \mathsf{C}(\boldsymbol{y}) &= E[\mathsf{G}\boldsymbol{x}(\mathsf{G}\boldsymbol{x})^{\mathrm{T}}] & (6.50)\\ &= E[\mathsf{G}\boldsymbol{x}\boldsymbol{x}^{\mathrm{T}}\mathsf{G}^{\mathrm{T}}] \\ &= \mathsf{G}E[\boldsymbol{x}\boldsymbol{x}^{\mathrm{T}}]\mathsf{G}^{\mathrm{T}} \\ \mathsf{C}(\boldsymbol{y}) &= \mathsf{G}\mathcal{C}(\boldsymbol{x})\mathsf{G}^{\mathrm{T}}. & (6.51) \end{aligned}$$

The last equality follows because the covariance matrix is symmetric. The expectation operator does not act on the elements of G because they are not random variables. The covariance matrix therefore transforms according to the similarity transformation (6.51).

The weighted average is a special case of linear transformation,

$$w = \sum_{i=1}^{M} a_i x_i = \boldsymbol{a}^{\mathrm{T}}\boldsymbol{x}, \tag{6.52}$$

where $\boldsymbol{a}$ is a vector of weights. Equation (6.51) applies with diagonal G, the diagonal elements being the weights a_i. The variance of W is therefore

$$\mathcal{V}[w] = \boldsymbol{a}^{\mathrm{T}}\mathsf{C}\boldsymbol{a}. \tag{6.53}$$

Another important case is when $\mathsf{G} = \mathsf{V}$, a matrix whose columns are the orthogonal eigenvectors of $\mathsf{C}(\boldsymbol{x})$. Then

$$\mathsf{V}^{\mathrm{T}}\mathsf{C}(\boldsymbol{x})\mathsf{V} = \Lambda, \tag{6.54}$$

where Λ is the diagonal matrix of eigenvalues of $\mathsf{C}(\boldsymbol{x})$ (see Appendix 4). Equation (6.51) shows that the transformed dataset $\boldsymbol{y}$ is independent (zero covariances or off-diagonal elements of the covariance matrix) and its variances are given by the eigenvalues of $\mathsf{C}(\boldsymbol{x})$. The dataset $\Lambda^{-\frac{1}{2}}\boldsymbol{y}$ is also *univariant*, the variances are all equal to one, because each component of $\boldsymbol{y}$ has been divided by its standard deviation. In general a data vector $\boldsymbol{x}$ may be transformed to an independent, univariant data vector by the transformation

$$\hat{\boldsymbol{x}} = \Lambda^{-\frac{1}{2}}\mathsf{V}^{\mathrm{T}}\boldsymbol{x}. \tag{6.55}$$

In the last section we saw how important it is to scale the individual equations of condition so that none dominate. The best procedure is to transform the data to univariant, independent form before performing the minimisation to obtain (6.25). Suppose the errors have zero mean and covariance matrix C_e. The sum of squares of the transformed univariant independent data is

$$E^2 = \boldsymbol{e}^T \mathsf{V} \Lambda^{-\frac{1}{2}} \Lambda^{-\frac{1}{2}} \mathsf{V}^T \boldsymbol{e} = \boldsymbol{e}^T \mathsf{C}_e^{-1} \boldsymbol{e}. \tag{6.56}$$

Setting $\boldsymbol{e} = \boldsymbol{d} - \mathsf{A}\boldsymbol{m}$ gives

$$\begin{aligned} E^2 &= (\boldsymbol{d} - \mathsf{A}\boldsymbol{m})^T \mathsf{C}_e^{-1} (\boldsymbol{d} - \mathsf{A}\boldsymbol{m}) \qquad (6.57) \\ &= \boldsymbol{d}^T \mathsf{C}_e^{-1} \boldsymbol{d} - 2\boldsymbol{d}^T \mathsf{C}_e^{-1} \mathsf{A}\boldsymbol{m} + \boldsymbol{m}^T \mathsf{A}^T \mathsf{C}_e^{-1} \mathsf{A}\boldsymbol{m}, \end{aligned}$$

where we have transposed scalars and used the symmetry of C_e^{-1}, for example

$$\boldsymbol{m}^T \mathsf{A}^T \mathsf{C}_e^{-1} \boldsymbol{d} = \boldsymbol{d}^T \mathsf{C}_e^{-1} \mathsf{A}\boldsymbol{m}.$$

Differentiating with respect to the elements of $\boldsymbol{m}$ gives

$$\frac{\partial E^2}{\partial \boldsymbol{m}} = -2\boldsymbol{d}^T \mathsf{C}_e^{-1} \mathsf{A} + 2\mathsf{A}^T \mathsf{C}_e^{-1} \mathsf{A}\boldsymbol{m} = 0, \tag{6.58}$$

which leaves

$$(\mathsf{A}^T \mathsf{C}_e^{-1} \mathsf{A})\boldsymbol{m} = \mathsf{A}^T \mathsf{C}_e^{-1} \boldsymbol{d}. \tag{6.59}$$

This solution differs from (6.25) by the appearance of C_e^{-1}. The effect of C_e^{-1} is to scale the equations properly; for independent data it divides each row and datum by the variances.

We end this section on a technical matter, on how to calculate the final misfit. The sum of squares of residuals is written in full as

$$E^2 = (\boldsymbol{d} - \mathsf{A}\boldsymbol{m})^T \mathsf{C}_e^{-1} (\boldsymbol{d} - \mathsf{A}\boldsymbol{m}). \tag{6.60}$$

E is properly called the *root mean square* (rms) misfit or residual. For a successful inverse the rms misfit should be about 1. A common measure of success of the model in fitting the data, particularly in tomographic problems, is to compare the original misfit, $\sqrt{\boldsymbol{d}^T \mathsf{C}_e^{-1} \boldsymbol{d}/D}$, with the final misfit, often as a percentage reduction in residual.

For small problems E is calculated by substituting the model back into the equations of condition, calculating $(\boldsymbol{d} - \mathsf{A}\boldsymbol{m})$ directly, but this usually involves storing or recalculating the big matrix A. An easier method is available. Rearranging (6.60) gives

$$E^2 = \boldsymbol{d}^T \mathsf{C}_e^{-1} \boldsymbol{d} + \boldsymbol{m}^T \mathsf{A}^T \mathsf{C}_e^{-1} \mathsf{A}\boldsymbol{m} - \boldsymbol{m}^T \mathsf{A}^T \mathsf{C}_e^{-1} \boldsymbol{d} - \boldsymbol{d}^T \mathsf{C}_e^{-1} \mathsf{A}\boldsymbol{m}. \tag{6.61}$$

The last two terms are equal: they can be transposed into each other. Furthermore, $\boldsymbol{m}$ satisfies the normal equations, which in the most general case are given by (6.59). The second term can therefore be replaced with

$$\boldsymbol{m}^T \mathsf{A}^T \mathsf{C}_e^{-1} \mathsf{A} \boldsymbol{m} = \boldsymbol{m}^T \mathsf{A}^T \mathsf{C}_e^{-1} \boldsymbol{d} \tag{6.62}$$

and (6.60) becomes

$$E^2 = \boldsymbol{d}^T \mathsf{C}_e^{-1} \boldsymbol{d} - \boldsymbol{m}^T \mathsf{A}^T \mathsf{C}_e^{-1} \boldsymbol{d}. \tag{6.63}$$

The first term is simply the sum of squares of weighted data, or original 'misfit'; the second is the scalar product of the model vector with the right-hand side of the normal equations (6.59). The scalar products are trivial to perform, and (6.63) provides a far more efficient method of calculating the rms misfit than (6.60) or (6.61).

6.4 Model covariance matrix and the error ellipsoid

The solution to (6.59) gives $\boldsymbol{m}$ as a linear combination of the data: it has the same form as (6.49) with

$$\mathsf{G} = \left(\mathsf{A}^T \mathsf{C}_e^{-1} \mathsf{A}\right)^{-1} \mathsf{A}^T \mathsf{C}_e^{-1}$$

We can therefore calculate the covariance matrix of the solution from (6.51):

$$\mathsf{C} = \left(\mathsf{A}^T \mathsf{C}_e^{-1} \mathsf{A}\right)^{-1} \mathsf{A}^T \mathsf{C}_e^{-1} \mathsf{C}_e \left[\left(\mathsf{A}^T \mathsf{C}_e^{-1} \mathsf{A}\right)^{-1} \mathsf{A}^T \mathsf{C}_e^{-1}\right]^T \tag{6.64}$$

$$= \left(\mathsf{A}^T \mathsf{C}_e^{-1} \mathsf{A}\right)^{-1} \left(\mathsf{A}^T \mathsf{C}_e^{-1} \mathsf{A}\right) \left(\mathsf{A}^T \mathsf{C}_e^{-1} \mathsf{A}\right)^{-1}$$

$$\mathsf{C} = \left(\mathsf{A}^T \mathsf{C}_e^{-1} \mathsf{A}\right)^{-1}. \tag{6.65}$$

C describes how the measurement errors have been mapped into the calculated model parameters. It can be used to assess the uncertainty in the components of $\boldsymbol{m}$. The covariance matrix can be used to define an *error ellipsoid*, sometimes called a confidence ellipsoid (see Appendix 4 for a discussion of the geometry of quadratic forms). Consider the random variables x and y with covariance matrix C_{xy}. We rotate these into independent variables ξ, η with standard deviations σ_ξ, σ_η calculated from the eigenvalues of C_{xy}. We may use the inequality

$$\frac{\xi^2}{\sigma_\xi^2} + \frac{\eta^2}{\sigma_\eta^2} \leq K^2 \tag{6.66}$$

to give a limit on the likely values of ξ and η, depending on the chosen value of the constant K. The equality in (6.66) defines an ellipse in the xy-plane with axes

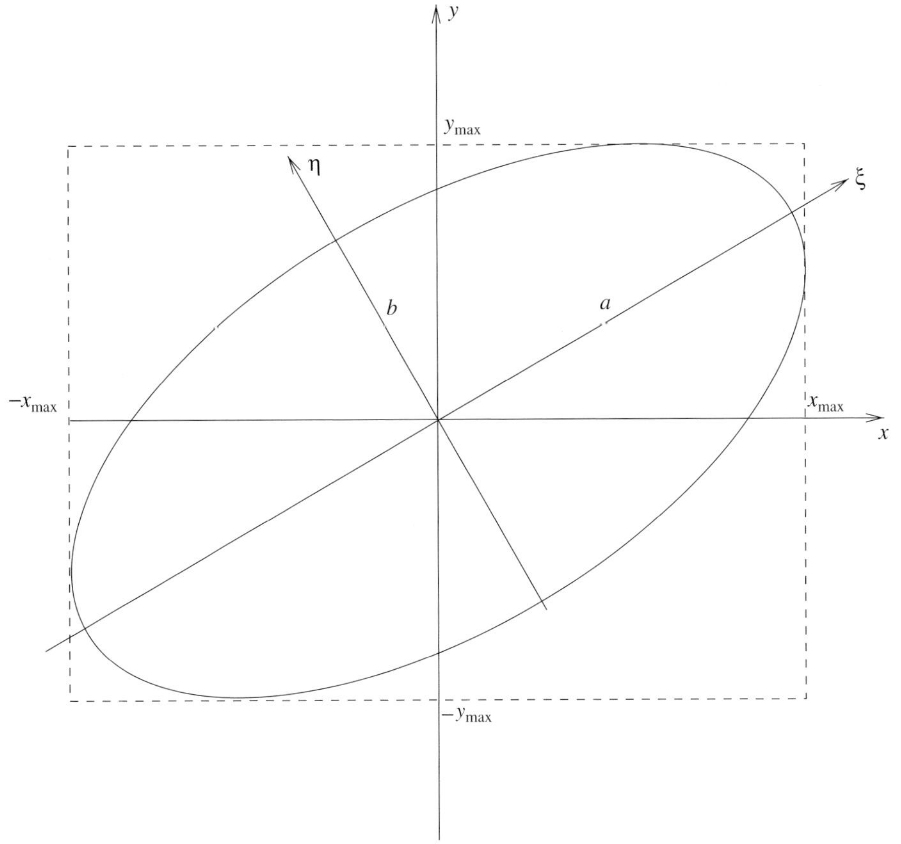

Fig. 6.4. The general error ellipse in two dimensions, showing the axes for the extrema in x and y.

along the eigenvectors of the covariance matrix C_{xy} (Figure 6.4). The semi-major and semi-minor axes are

$$a = \frac{\sigma_\xi}{K} = \frac{\lambda_1}{K} \tag{6.67}$$

$$b = \frac{\sigma_\eta}{K} = \frac{\lambda_2}{K} \tag{6.68}$$

where λ_1, λ_2 are the eigenvalues of the covariance matrix. K may be chosen to make the ellipse the 95% confidence level if enough is known about the statistics of the error, or it may be chosen less rigorously. The theory carries over into many dimensions. The geometrical form is a (hyper)ellipsoid because the eigenvalues are all positive (Appendix 4).

Figure 6.4 also shows how to estimate extreme values of x and y graphically. Analytically, we wish to find the maximum value of x and y for a point constrained to be on the bounding ellipse. The method works in any number of dimensions. The equation of the ellipse in vector form is $\boldsymbol{x}^T \mathsf{C}_e^{-1} \boldsymbol{x} = K^2$, where $\boldsymbol{x} = (x, y)$ is the position vector. Extreme values of x and y are found using the method of Lagrange multipliers (Appendix 6) and minimising the two quantities

$$x - l_x \left(\boldsymbol{x}^T \mathsf{C}_e^{-1} \boldsymbol{x} - K^2 \right) \tag{6.69}$$

$$y - l_y \left(\boldsymbol{x}^T \mathsf{C}_e^{-1} \boldsymbol{x} - K^2 \right), \tag{6.70}$$

where l_x, l_y are Lagrange multipliers. Differentiating each expression with respect to x and y and setting the results to zero gives

$$2 l_x \mathsf{C}_e^{-1} \boldsymbol{x} = \begin{pmatrix} 1 \\ 0 \end{pmatrix} \tag{6.71}$$

$$2 l_y \mathsf{C}_e^{-1} \boldsymbol{x} = \begin{pmatrix} 0 \\ 1 \end{pmatrix}. \tag{6.72}$$

The solution of the first pair of equations gives the direction of the axis through the maximum and minimum x-values; the second the corresponding axis through the maximum and minimum of the y-values. The Lagrange multipliers are found from the constraint $\boldsymbol{x}^T \mathsf{C}_e \boldsymbol{x} = K^2$. In practice we would drop the constants l_x, l_y from equations (6.71) and (6.72) and solve for the required directions, then normalise them to bring them onto the ellipse. The required extreme values for x and y are then the components of this vector.

We apply these ideas to problem (4) in Table 6.2 for the density of the Earth. This is the simplest problem that allows calculation of a non-trivial covariance matrix because the others yield only a scalar uncertainty on $\bar{\rho}$. The data M and I/a^2 are independent and their assigned errors are the same, $\pm 0.003 \times 10^{24}$ kg. The covariance matrix of the data C_e is therefore 9×10^{-18} times the 2×2 identity matrix. Substituting (6.18) for A into (6.65) for the model covariance matrix gives

$$\mathsf{C} = \left(\mathsf{A}^T \mathsf{C}_e^{-1} \mathsf{A} \right)^{-1} = 9 \times 10^{42} \left(\mathsf{A}^T \mathsf{A} \right)^{-1} = \begin{pmatrix} 3076 & -526 \\ -526 & 99 \end{pmatrix}. \tag{6.73}$$

The eigenvalues are 8.828, 3166. The corresponding normalised eigenvectors are (0.1690, 0.9856) and (0.9856, −0.1690). The best-determined combination is therefore $0.1690\rho_c + 0.9856\rho_m$, corresponding to the minor axis of the ellipse in Figure 6.5. The best-determined combination is close to ρ_m; its error is 3 kg m^{-3} (the square root of the eigenvalue). The poorest-determined combination is ρ_c with an error of 56 kg m^{-3}. As before, these error estimates are absurdly low because of the overly simple two-layer Earth model.

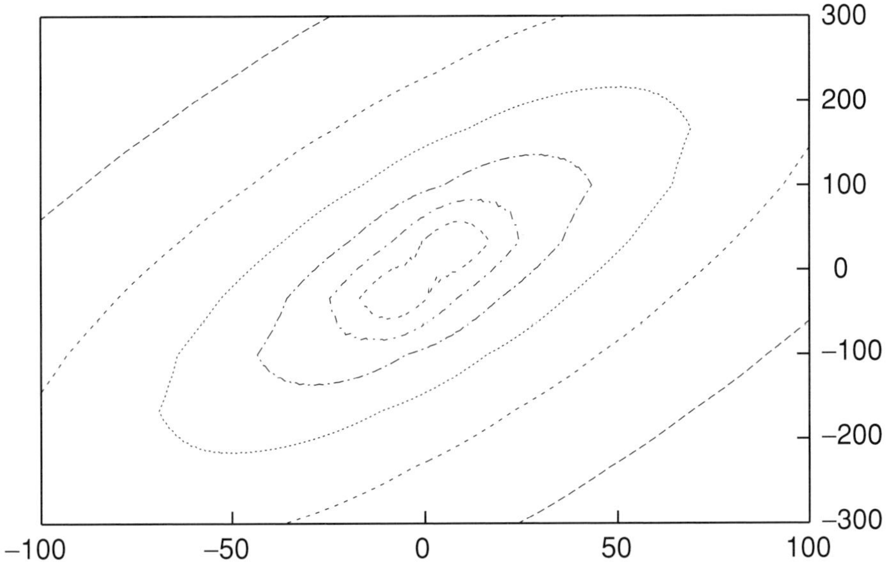

Fig. 6.5. Error ellipses of the mantle and core density parameters obtained from the covariance matrix in equation (6.73). Contours are logarithmic scale.

6.5 Robust methods

However careful we are in weighting our data, some mistakes will remain. These present a huge problem for the least-squares method because the squaring process increases their influence on the solution. A mistake leading to a datum that is effectively wrong by $+10$ standard deviations will have the same effect as 100 data biassed by just $+1$ standard deviation, or 25 data biassed by $+2$ standard deviations. In practice something must be done to remove these 'outliers', or reduce their effect to a minimum. Methods designed to be insensitive to outliers are collectively called *robust* methods.

The simplest solution is to calculate the residual (datum minus model prediction) and throw away any data with residuals exceeding some threshold value, usually 3–5 standard deviations. The method of data rejection has several advantages: it usually works, is easy to implement, and requires little thought. There is, in fact, little alternative when faced with data that lie very far from the model prediction, but better methods have been devised for more subtle difficulties. The least-squares method is derived by minimising a Gaussian likelihood function (pp. 97–98) and is only valid when the law of errors for the data follows a Gaussian curve. Problems can be identified by plotting residuals on a histogram; the histogram should have the approximate bell-shape of a Gaussian function. All too often the histogram shows too many residuals away from the main central peak.

In the early days of seismology Harold Jeffreys found prominent outliers in seismic arrival time data. These are particularly evident for second arrivals, such as S waves, that have to be identified and read at a time when the seismogram is already disturbed. This can increase the scatter in the readings by a factor of 10, violating the Gaussian assumption. The histogram of residuals approximates a Gaussian curve in the centre but falls off less rapidly at the extremes. He therefore developed the *method of uniform reduction*, which is based on a law of errors giving the probability $p(x)$ of residual x as a linear combination of a Gaussian and a linear function, with the linear function being small relative to the Gaussian. This reproduces the shape of the distribution of seismic travel times. The maximum likelihood solution is developed rigorously in Jeffreys' classic monograph on the Theory of Probability (pp. 214ff), where he also criticises the theoretical basis for data rejection. Implementation of the method involves simple downweighting of outliers. The method of uniform reduction was adopted by the International Seismological Service in location of earthquakes, and is still used today by its successor organisation the International Seismological Centre. Elsewhere the method has been abandoned; I do not know if this is because modern seismologists have found better methods, have forgotten about it, or never knew about it.

Another approach is to change the law of errors from Gaussian to some other probability distribution. We shall consider distributions of the form

$$p(x) = K e^{|x - x_L|^p} \tag{6.74}$$

with $p \geq 1$. The case $p = 2$ gives the usual Gaussian distribution. The other important case is $p = 1$, the Laplace or double-exponential distribution. The derivation of the maximum likelihood solution given on (p. 97) proceeds as before. Taking $p = 1$ clearly downweights outliers relative to larger values of p. A single datum with a residual of $+10$ standard deviations now has the influence of only 10 data with residuals of $+1$ standard deviations, rather than 100 as in least squares. Outliers become progressively more important for solutions as p increases, until $p \to \infty$, when only the maximum residual influences the solution.

The modulus in (6.74) makes the likelihood function nonlinear, and some algorithms for finding the solution are very clumsy. Fortunately, a very simple iterative procedure seems to work well in many geophysical examples. For univariate, uncorrelated, data the maximum likelihood solution involves minimising the quantity

$$E^p = \sum_{i=1}^{D} \left| d_i - \sum_{j=1}^{P} A_{ij} m_j \right|^p . \tag{6.75}$$

E is usually called the $\mathcal{L}^p$-norm of residuals. Differentiating with respect to the ms and setting the result to zero in the usual way leads to the equations

$$\sum_{k=1}^{P}\sum_{i=1}^{D}\left(\frac{A_{ik}d_k}{r_i^{2-p}} - \frac{\sum_{j=1}^{P} A_{ji}A_{jk}m_k}{r_i^{2-p}}\right) = 0, \tag{6.76}$$

where r_i is the residual of the ith datum

$$r_i = d_i - \sum_{j=1}^{P} A_{ij}m_j. \tag{6.77}$$

Equations (6.76) may be put into matrix form by defining a weight matrix W with elements

$$W_{ij} = \frac{\delta_{ij}}{|r_i|^{p-2}} \tag{6.78}$$

to give

$$\mathsf{A}^{\mathrm{T}}\mathsf{W}\mathsf{A}\boldsymbol{m} = \mathsf{A}^{\mathrm{T}}\mathsf{W}\boldsymbol{d}. \tag{6.79}$$

The matrix multiplying $\boldsymbol{m}$ on the left-hand side is square, positive definite, and invertible. Note that the least-squares case has $p = 2$, when this equation reduces to the standard least-squares solution for univariate uncorrelated data (6.25).

Comparing equation (6.79) with (6.59) shows them to be the same with data weighted with W rather than C_e^{-1}. Equation (6.79) cannot be solved immediately because W depends on the model $\boldsymbol{m}$ through (6.77). The solution can be obtained iteratively. Start from some initial model $\boldsymbol{m}_0$, perhaps by setting $\mathsf{W}_0 = \mathsf{I}$. Calculate some initial residuals from (6.77) and the corresponding weight matrix W_1. Solve for a new model $\boldsymbol{m}_1$ using (6.79) and calculate the new residuals and weight matrix W_2, and start again. Iteration proceeds to convergence. This method is called *iterative reweighted least squares* (IRLS) because each step consists of a least squares solution with weights determined from the residuals at the previous step.

The iterations are guaranteed to converge for $p \geq 1$ provided certain simple conditions are met. The most important of these arises when one of the residuals happens to lie close to zero, or is so small that it causes numerical problems in the calculation. This difficulty is easily resolved by adding a small positive number ϵ to the residual, replacing (6.78) with

$$W_{ij} = \frac{\delta_{ij}}{|r_i|^{2-p} + \epsilon}. \tag{6.80}$$

A seismological example of IRLS is given by Gersztenkorn *et al.* (1986), who inverted the 1D acoustic approximation using the Born approximation. A geomagnetic example is given by Walker and Jackson (2000), who find a model

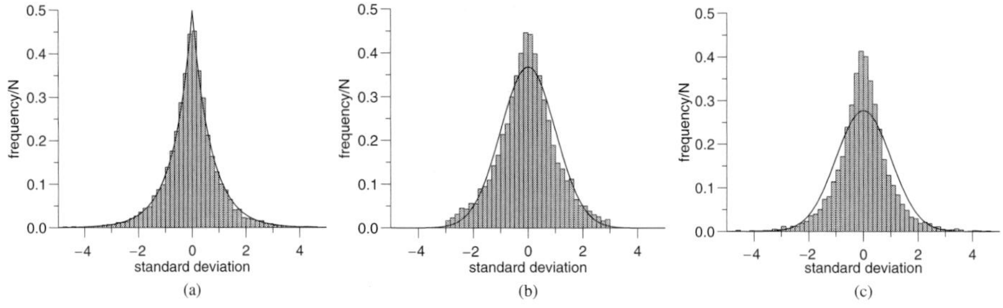

Fig. 6.6. Histograms of residuals of geomagnetic measurements from a model of the large scale core field. (a) Residuals from an L^1-norm model compared with the double-exponential distribution (solid line); (b) residuals from a conventional least squares model without data rejection compared with the Gaussian distribution; and (c) least squares with data rejection. Reproduced, with permission, from Walker and Jackson (2000).

of the Earth's main magnetic field (see also Chapter 12). Magnetisation of crustal rocks presents a major source of error in the form of small scale fields, and there is good evidence that the crustal field has a Laplace distribution, as shown by the residuals in Figure 6.6. An L^1 norm applied by IRLS gives a final histogram of residuals that fits the ideal Laplace distribution well, whereas conventional least squares ($p = 2$) gives a rather poor fit to a Gaussian distribution, even with rejection of outliers. A recipe for linear parameter estimation is given in Box 6.1.

Box 6.1. Summary: recipe for linear parameter estimation

(i) Establish the form of the linear relationship between the data and model parameters. Define the data vector and the model vector, and the elements of the equation of condition matrix A.

(ii) Assign an error estimate to every datum. The error estimate is an inseparable part of the measurement.

(iii) Plot histograms of the errors to establish their distribution.

(iv) Form the error covariance matrix C_e^{-1}. If the data vector comprises combination of original, independent, measurements, the error estimates must also be transformed appropriately.

(v) Solve the normal equations, either by transforming to independent, univariant, data and using (6.26) or by using the data covariance matrix in (6.59).

(vi) Compute the residuals from the model predictions.

(vii) Examine the residuals individually (if possible) and examine their distributions. Ideally the weighted residuals should be unity. Examine any excessively large residuals for blunders. Remove outliers, or downweight them, and compute a new set of model parameters.

(viii) Compute the model covariance matrix.

EXERCISES

6.1. Verify equation (6.37).

6.2. (a) Find the eigenvalues and normalised eigenvectors of the matrix

$$\mathsf{A} = \begin{pmatrix} 5 & 0 & \sqrt{3} \\ 0 & 1 & 0 \\ \sqrt{3} & 0 & 3 \end{pmatrix}.$$

(b) Verify the eigenvectors are orthogonal by taking scalar products.
(c) Form the matrix V from the eigenvectors and show it is orthogonal.
(d) Verify that $\mathsf{V}^T A \mathsf{V}$ is diagonal with eigenvalues along the diagonal.
(e) Sketch the quadratic form $\boldsymbol{x}^T \mathsf{A} \boldsymbol{x}$, indicating its principal axes of symmetry.

6.3. Fit a straight line $y = ax + b$ through the three data points $(-1, 0.1)$; $(0, 1.0)$; $(1, 1.9)$ by least squares using the following method.

(a) Write down the coefficients of the equations of condition matrix A.
(b) Evaluate the normal equations matrix $\mathsf{A}^T\mathsf{A}$.
(c) Evaluate $\mathsf{A}^T\boldsymbol{d}$.
(d) Write down the normal equations and solve for the model vector (a, b). Verify that, in this case, the least-squares solution is the exact solution.
(e) Change the last point to (1.0, 1.8) and recalculate the best-fitting straight line.
(f) Plot your points and straight lines on graph paper.

6.4. A pair of random variables X, Y are sampled three times to give the values (1.2, 3.7); (1.5, 5.2); and (1.6, 5.8).

(a) Estimate the means and standard deviations of X and Y.
(b) Estimate the elements of the covariance matrix $\mathsf{C} = \mathcal{C}[X, Y]$.
(c) Estimate $\mathcal{V}[X + Y]$ directly from the three measurements of $X + Y$.
(d) Compare this estimate with the variance obtained from $\mathcal{V}[X] + 2\mathcal{C}[X, Y] + \mathcal{V}[Y]$.

6.5. The epicentre of an earthquake is located by a seismic array. The location procedure gives a covariance matrix of the epicentral coordinates (X, Y) (referred to cartesian axes with origin at the epicentre, so $E[X] = E[Y] = 0$):

$$\mathsf{C} = \begin{pmatrix} 2 & 1 \\ 1 & 2 \end{pmatrix}. \qquad \text{(E6.1)}$$

(the units are km^2).
$\boldsymbol{X} = (x, y)$ is a position vector in the xy-plane.

(a) Find the equation of the error ellipsoid of the epicentre in terms of x and y.
(b) Sketch the error ellipsoid (by finding eigenvalues and eigenvectors of C) and find the direction of maximum error and the error in that direction.

6.6. **Computer exercise: introduction to `octave` and `gnuplot`**

`octave` is a Free Software Foundation program that may be obtained over the web (see Appendix 7). It is similar to the commercial product Matlab. It can be used to perform most matrix manipulations with simple on-line commands. It is also possible to write scripts and programs in `octave`. It is best to do these computer exercises by building up the commands in a separate file and running the file; this leaves a record of what you have done.

`octave` uses `gnuplot`, the Free Software Foundation's graphics package, to do the plotting. `gnuplot` can also be obtained over the web (see Appendix 7).

(a) Use `octave` to verify your solutions to questions 6.2(a)–(c).
(b) Find the inverse of the matrix in question 6.1 by constructing $\mathsf{V}\Lambda^{-1}\mathsf{V}^{\mathrm{T}}$, where Λ has its usual meaning of the diagonal matrix of eigenvalues. Verify it is the inverse by multiplying with the original matrix to obtain the identity. Can you prove that?
(c) Check your solution to question 6.3.
(d) Use `gnuplot` (Appendix 7) to produce a plot of the straight line obtained for question 6.3, including points for the original data, and compare it with your hand-drawn solution.

6.7. **Computer exercise**

Do a least-squares fit of a quadratic $y = ax^2 + bx + c$ to the following data

x:	0	1	2	3	4
y:	0.1	1.2	2.7	6.9	15.0

by the following method.

(a) Form the equations-of-condition matrix A; 5 by 3 matrix.
(b) Form the normal equations matrix $\mathsf{A}^{\mathrm{T}}\mathsf{A}$.
(c) Form the vector $\mathsf{A}^{\mathrm{T}}\boldsymbol{y}$.
(d) Obtain the least-squares solution $\boldsymbol{m} = (\mathsf{A}^{\mathrm{T}}\mathsf{A})^{-1}\mathsf{A}^{\mathrm{T}}\boldsymbol{y}$.
(e) Plot the quadratic $y(x)$ as a continuous line and the data on an xy-plot using `gnuplot` as described below.

6.8. **Computer exercise: weighting data**

The purpose of this practical is to learn how to treat data with widely different accuracy within the same inversion and identify bad data. Ideally you should have an error estimate on every measurement, but mistakes can often occur that can ruin the inversion.

I generated data by the formula $y = ax + b$, added Gaussian noise, then changed three of the y-values. You have to identify the three bad values and remove them, then obtain the best estimates of a and b. The data are in file `prac3.data`.

(a) Find a least-squares straight-line fit to the data. Plot the data and your line together.

(b) How many obviously bad data can you see?

(c) Remove them, but only if you are absolutely sure they are bad. Find a new straight line fit and plot it against the data.

(d) Can you see any more obviously bad data? If so, remove them. Now weight the data: in adding noise I used normally distributed numbers with a standard deviation $\sigma = 1, 2, 1, 4$ repeated in cycles. Construct a diagonal weight matrix C_e^{-1} from the third column of `xys`, which contains the variances. Assume the data are independent. Invert for a straight line again. Plot it against the data.

(e) Now can you see any bad data? If not, construct the weighted residual vector $\mathsf{C}_e^{-1/2}(\boldsymbol{d} - \mathsf{A}\boldsymbol{m})$ and look for any residuals significantly greater than 1. Remove the last bad datum and do one final weighted inversion.

(I used $a = 2, b = 5$.)

7

The under-determined problem

7.1 The null space

Now consider the case when there are fewer linearly independent equations (6.4) than unknowns. Even in the absence of errors it is impossible to find a unique solution. There exists at least one model $\boldsymbol{m}_0$, called a *null vector*, such that

$$\mathsf{A}\boldsymbol{m}_0 = 0. \tag{7.1}$$

This arises if $D < P$ or when $D \geq P$ but $D - P$ or more of the equations are linear combinations of the rest, as would happen for example if we repeatedly measured the same thing in order to reduce error on the mean. A solution $\boldsymbol{m}_1$ cannot be unique because

$$\boldsymbol{m} = \boldsymbol{m}_1 + \alpha \boldsymbol{m}_0 \tag{7.2}$$

will also be a solution, where α is any scalar multiplier. The space spanned by all null vectors is called the *null space*; it has dimension $N \geq D - P$.

It is easy to understand the physical meaning of the null space. Matrix A contains the rule for calculating the measurement from the model; it embodies the physical laws used in the forward problem. A null vector is a model that produces zero measurement. It produces nothing measurable, and we cannot therefore expect to understand anything about that particular model from the measurements. For a linear problem these unknown models can be added to any solution that fits the data to produce another solution that fits the data equally well. Attempting to solve the normal equations (6.25) when A has a null space would result in failure because $\mathsf{A}^T\mathsf{A}\boldsymbol{m}_0$ is also zero. The normal-equations matrix is singular and non-invertible.

Problems (5) and (6) of Table 6.2 (p. 88) illustrate the idea of a null space. Equation (6.16) applies for problem (5). There is one equation for the two unknowns ρ_c and ρ_m, defining the straight line AB on a graph of ρ_m against ρ_c (Figure 7.1).

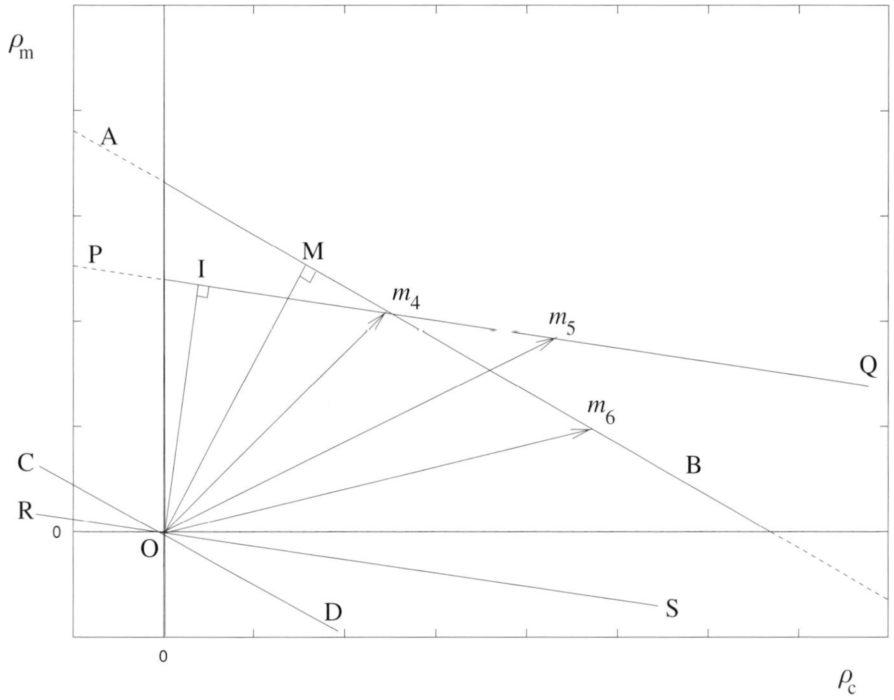

Fig. 7.1. The model space for the problem of determination of density for a two-layer Earth using one of M or I. Not drawn to scale.

Any vector like $\boldsymbol{m}_6$, with its arrow tip on line AB, is an allowed solution, one that fits the datum exactly.

The null space for problem (5) is given by the equation (6.16) with zero replacing M:

$$c^3 \rho_c + (a^3 - c^3)\rho_m = 0. \tag{7.3}$$

This is another line, CD in Figure 7.1, parallel to AB and passing through the origin. The general solution in vector form is

$$\boldsymbol{m} = \vec{\mathrm{OM}} + \alpha \vec{\mathrm{CD}}, \tag{7.4}$$

a graphical representation of (7.2). The same analysis applies to problem (6) using only I. Equation (6.17) defines the straight line PQ in Figure 7.1, the null space defines the line RS, and $\boldsymbol{m}_5$ is a solution. The point of intersection of lines AB and PQ satisfies both data and $\boldsymbol{m}_4$ is the unique solution of problem (4).

In this particular example we have additional constraints: the densities must be positive. The inverse problem, as posed, does not know about this positivity constraint and allows solutions with negative density (lying on the dashed parts of the lines AB and PQ in Figure 7.1).

The ideas of model space and null space carry over into many dimensions and give a useful language with which to describe the full family of solutions. A single measurement for this problem reduces the dimension of the space of allowed solutions by 1. It may not produce a unique solution but nor is it a useless measurement.

Mass and moment of inertia were used to good practical effect by Bullen (1975) to eliminate a whole class of density models for the Earth. The models were based on seismic velocities and the Adams–Williamson equation, and many produced densities that gave $I > 0.4Ma^2$, the value for a uniform sphere. This entailed denser material overlying lighter material, which is gravitationally unstable and therefore physically impossible. Incorporating M and I into the dataset resolved the larger ambiguities.

7.2 The minimum-norm solution

Satisfying though it is to describe the full solution to the inverse problem, sooner or later we need to plot a model that fits the data. This requires selection, a commitment that should be based on additional information but often we can use only our personal taste and prejudice. A useful choice is the shortest solution, mathematically defined as the one with least length or norm, $|\boldsymbol{m}|$. Any model in the null space can be added to this solution without changing the fit to the data, E. It is therefore not an essential part of the model and can be omitted. The minimum-norm solution is therefore the one orthogonal to the null space. In Figure 7.1, this is model $\vec{\mathrm{OM}}$, the perpendicular from the origin to line AB for problem (5). For problem (6) it is $\vec{\mathrm{OI}}$, the perpendicular to line CD.

The *minimum-norm solution* contains no unnecessary information, which is the philosophy justifying its popularity. Minimising $\boldsymbol{m}^{\mathrm{T}}\boldsymbol{m}$ subject to the constraint $\mathsf{A}\boldsymbol{m} = \boldsymbol{d}$ (equation (6.4) without errors) gives the solution (see Appendix 4, equation (A4.43))

$$\boldsymbol{m} = \mathsf{A}^{\mathrm{T}}(\mathsf{A}\mathsf{A}^{\mathrm{T}})^{-1}\boldsymbol{d}. \tag{7.5}$$

Compare this with the least-squares solution (6.26). The square matrix $\mathsf{A}\mathsf{A}^{\mathrm{T}}$ is $D \times D$ whereas the normal-equations matrix $\mathsf{A}^{\mathrm{T}}\mathsf{A}$ is $P \times P$. Only the square matrix with lower dimension has an inverse: (6.26) works when $D > P$, (7.5) works when $P > D$. Each equation may be thought of as defining a generalised inverse of the rectangular matrix A:

$$\mathsf{A}^{-N} = (\mathsf{A}^{\mathrm{T}}\mathsf{A})^{-1}\mathsf{A}^{\mathrm{T}} \tag{7.6}$$

$$\mathsf{A}^{-M} = \mathsf{A}^{\mathrm{T}}(\mathsf{A}\mathsf{A}^{\mathrm{T}})^{-1} \tag{7.7}$$

$$\mathsf{A}^{-N}\mathsf{A} = \mathsf{A}\mathsf{A}^{-M} = \mathsf{I}. \tag{7.8}$$

A^{-N} and A^{-M} can be called the left and right inverses of A respectively.

Broadening the definition of the norm can make selection of a solution more flexible. For example, to find the solution fitting the data that is also closest to some preconceived model $\boldsymbol{m}_{\mathrm{P}}$, we can minimise the norm $|\boldsymbol{m} - \boldsymbol{m}_{\mathrm{P}}|$. Minimising the square of a derivative can be useful in producing a smooth solution. The solution norm is generalised to the form

$$\boldsymbol{m}^{\mathrm{T}}\mathsf{W}\boldsymbol{m}, \tag{7.9}$$

where W can be any symmetric, positive definite matrix of weights. W is the identity for the usual norm $\boldsymbol{m}^{\mathrm{T}}\boldsymbol{m}$; the more general form allows some model parameters to be weighted more heavily than others (for example, we might believe, on the basis of information other than the data being inverted, that they are smaller), in addition to allowing for correlations between parameters. W must be positive definite or certain combinations of parameters could be amplified without limit. The formula $|\boldsymbol{m}| < K$ defines an ellipsoid in P-dimensional model space; the solution is therefore confined to a closed volume by placing an upper bound on its norm. The data constraints (6.4) describe a plane in model space because they are a set of linear equations. The minimum norm solution then lies at the point nearest the origin on the intersection of the data plane and ellipsoid.

7.3 Ranking and winnowing

Eigenvector expansions give an alternative way of solving the normal equations (6.25) when A has null vectors. The normal equations matrix $\mathsf{A}^{\mathrm{T}}\mathsf{A}$ is square and symmetric; its eigenvectors span model space and are mutually orthogonal. The eigenvectors may be used as a basis, and any model may be expanded as a linear combination of them.

Multiplying (6.4) by A^{T} gives

$$\mathsf{A}^{\mathrm{T}}\mathsf{A}\boldsymbol{m} = \mathsf{A}^{\mathrm{T}}(\boldsymbol{d} - \boldsymbol{e}). \tag{7.10}$$

Let the orthonormal eigenvectors and corresponding eigenvalues of $\mathsf{A}^{\mathrm{T}}\mathsf{A}$ be $\boldsymbol{u}^{(i)}$ and $\lambda^{(i)}$:

$$\mathsf{A}^{\mathrm{T}}\mathsf{A}\boldsymbol{u}^{(i)} = \lambda^{(i)}\boldsymbol{u}^{(i)}, \tag{7.11}$$

where i runs from 1 to P. Expanding $\boldsymbol{m}$ and the right-hand side of (7.10) in eigenvectors,

$$\boldsymbol{m} = \sum_{i=1}^{P} m^{(i)}\boldsymbol{u}^{(i)}, \tag{7.12}$$

and substituting into (7.10) gives

$$m^{(i)} = \frac{\boldsymbol{u}^{(i)^{\mathrm{T}}} \mathsf{A}^{\mathrm{T}} (\boldsymbol{d} - \boldsymbol{e})}{\lambda^{(i)}}. \tag{7.13}$$

In the absence of error, a zero eigenvalue $\lambda^{(i)} = 0$ leads to $0/0$ in one term of the right-hand side, which is indeterminate. Setting the coefficient $m^{(i)}$ to zero removes this indeterminate term and gives a solution to the normal equations that is orthogonal to the null space. It is therefore the shortest possible solution, the minimum-norm solution, the same as was obtained from (7.5).

Consider problem (5), inverting M for two densities. The normal-equations matrix is 2×2. One eigenvector obviously corresponds to zero eigenvalue; it points in the direction of line CD in Figure 7.1. The other eigenvector must be orthogonal to the first, parallel to line OM in Figure 7.1, the minimum-norm solution.

When an eigenvalue is small but non-zero the factor $1/\lambda^{(i)}$ tends to produce a large but determinate contribution on the right-hand side of (7.12). With perfect, error-free data that are consistent with the model the numerator in (7.13) will also be small and the ratio produces no difficulty, but error has no respect for consistency and can generate a large contribution. Thus although $|\boldsymbol{d}| \gg |\boldsymbol{e}|$, there may be cases where

$$\boldsymbol{u}^{(i)\mathrm{T}} \mathsf{A}^{\mathrm{T}} \boldsymbol{d} \ll \boldsymbol{u}^{(i)\mathrm{T}} \mathsf{A}^{\mathrm{T}} \boldsymbol{e}. \tag{7.14}$$

The solution is then dominated by the error, which is amplified by the smallness of the eigenvalue.

The null eigenvectors represent models that contribute nothing to satisfying the data. The remaining eigenvalues can be ranked in order of their size (they are all positive). The eigenvector corresponding to the smallest (non-zero) eigenvalue is a model contributing least to satisfying the data; that corresponding to the largest eigenvalue contributes most. In the presence of errors some of these small eigenvalues will be indistinguishable from zero and their eigenvectors should therefore be treated as part of the null space.

The eigenvalues are therefore separated into three sets: those that are genuinely zero because of the nature of the measurements, those that are so small as to contribute very little to the solution by way of reducing E, and those that make an important contribution. The generalised matrix inverse omits eigenvectors with eigenvalues that are identically zero; it is derived in Appendix 4, equation (A4.36). It may be extended by omitting also those eigenvectors with eigenvalues that fall below some threshold that depends on the size of the errors. This removes that part of the model most prone to error. It is said to *stabilise* the solution. Choice of threshold is nearly always subjective; it is made easier if the spectrum of eigenvalues has a clear break between large and small.

Ordering the eigenvalues by size is called *ranking*; removing the small eigenvalues is called *winnowing*. An advantage of this procedure is to reduce the dimension of model space down to the minimum required by the existing data before doing further processing.

7.4 Damping and the trade-off curve

A third method of stabilising an under-determined problem is to replace the rigid minimum norm criterion with a more flexible criterion that minimises a combination of error E and solution norm N,

$$T(\theta) = E + \theta^2 N. \tag{7.15}$$

θ is called a *trade-off parameter*; it determines the relative importance of E and N and is squared only to emphasise that it remains positive. Minimising T is a compromise between minimising E, fitting the data better, or minimising N, giving a 'simple' model, depending on the choice of θ. If $\theta = 0$ the solution reduces to (6.26), the conventional solution of the normal equations. When $\theta = \infty$ the data are ignored completely and the norm is minimised, with the result $\boldsymbol{m} = 0$ in the absence of further constraints. A plot of E versus N for different values of θ is called the *trade-off curve*, which has the shape idealised in Figure 7.2. $T(\theta)$ is monotonic and asymptotes to the minimum value of E (which may be zero for an under-determined problem) at one end and the minimum value allowed for the model norm N at the other.

The minimisation may be performed by differentiation along the same lines as the derivation of the normal equations (6.59). Weighting the data with its covariance matrix $\mathsf{C_e}$ gives

$$(\mathsf{A}^T\mathsf{C}_e^{-1}\mathsf{A} + \theta^2\mathsf{W})\boldsymbol{m} = \mathsf{A}^T\mathsf{C}_e^{-1}\boldsymbol{d}. \tag{7.16}$$

This equation differs from (6.59) only by the addition of the weight matrix W on the left-hand side. This method of stabilising the inversion is called *damping*.

In the important special case $\mathsf{W} = \mathsf{I}$ the difference amounts to adding a constant to each diagonal element of the normal equations matrix. The eigenvalues of $(\mathsf{A}^T\mathsf{A} + \theta^2 i\mathsf{I})$ are $(\lambda^{(i)} + \theta^2)$. Damping with a unit weight matrix has the effect of increasing all eigenvalues by a constant value. If $\lambda^{(i)} \gg \theta^2$ the model remains relatively unchanged, but if $\lambda^{(i)} \leq \theta^2$ the contribution to the eigenvector expansion (7.12) is reduced. Contributions from noisy model vectors is therefore reduced by choosing a value of θ^2 that is appropriate for the noise level in the data. The effect is similar to winnowing, but the sharp cut-off of the eigenvector expansion imposed by winnowing is replaced by a gradual reduction in amplitude of the noisier eigenvectors.

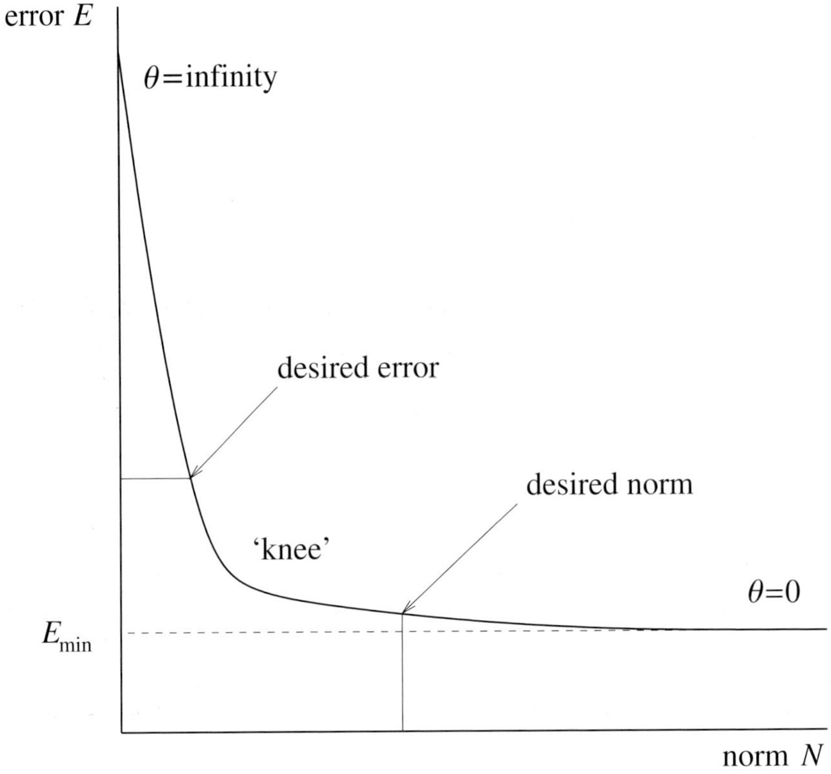

Fig. 7.2. The trade-off curve $T(\theta)$ gives the misfit E against the model norm N, a measure of the model's complexity. Each point on the curve represents a solution, allowing us to balance the compromise between fitting the data well and obtaining a simple model.

Choosing a suitable value of θ is somewhat *ad hoc*, although there are some optimal criteria. There are three different philosophies with essentially the same algebra.

Mapping. Here we just want a nice model for interpretation purposes; we adjust the damping parameter until we get the desired effect.

Data fitting. If we know the accuracy of the data well we adjust θ until the misfit to the weighted data is unity.

Compromise. If we have no confidence in either the model complexity or the data error, the knee of the curve (Figure 7.2) is a compromise.

Problem (5), to find the densities of core and mantle from the Earth's mass, can be used to illustrate the effect of damping on a very simple problem. We first simplify (6.16) to the standard form $d = \mathsf{A}\boldsymbol{m}$ by defining

$$d = \frac{3M}{4\pi a^3}; \quad x = \left(\frac{c}{a}\right)^3 = 0.163\,68, \tag{7.17}$$

which gives

$$d = x\rho_c + (1 - x)\rho_m. \tag{7.18}$$

Matrix A is $(x, 1 - x)$, so

$$\mathsf{A}^T\mathsf{A} = \begin{pmatrix} 0.026\,790 & 0.136\,886 \\ 0.136\,886 & 0.699\,437 \end{pmatrix}.$$

We can now construct a trade-off curve by solving the equation $(\mathsf{A}^T\mathsf{A} + \theta^2\mathsf{I})\boldsymbol{m} = \mathsf{A}^T d$ repeatedly for different values of the trade-off parameter θ. When θ is small, 10^{-3}, we recover the solution $(\rho_c, \rho_m) = (1243.0, 6351.2)$ kg m^{-3}, which is the same as the minimum-norm solution to the accuracy given. This is to be expected because even weak damping can remove the effect of the zero eigenvalue of $\mathsf{A}^T\mathsf{A}$ that corresponds to the null space. Increasing θ gradually reduces the norm while impairing the fit to the data. In this case the trade-off curve is a straight line; there is no knee to indicate an optimal choice of damping constant. This is a consequence of the simplicity of this particular problem: the eigenvector corresponding to the null space has no effect after a small amount of damping, leaving a 1D problem. Damping a 1D problem simply increases the misfit in proportion to the reduction in the norm of the model, until eventually the model is zero and the misfit equal to the original datum.

All solutions along the trade-off curve, including the minimum-norm solution, have $\rho_m > \rho_c$. This follows directly from (7.18), in which the coefficient of ρ_m greatly exceeds that multiplying ρ_c. Minimising the sum of squares of ρ_c and ρ_m while satisfying (7.18) leads to larger values of ρ_m: the penalty paid in the minimisation is smaller than making ρ_c large. These solutions have $\rho_m > \rho_c$, which is physically implausible because a planet with an exterior denser than its interior is gravitationally unstable. A more plausible solution is discussed in the next section.

7.5 Parameter covariance matrix

The solution of (7.16) is formally

$$\boldsymbol{m} = \mathsf{H}\boldsymbol{d}, \tag{7.19}$$

where the matrix H is $P \times D$:

$$\mathsf{H} = \left(\mathsf{A}^T\mathsf{C}_e^{-1}\mathsf{A} + \theta^2\mathsf{W}\right)^{-1} \cdot \mathsf{A}^T\mathsf{C}_e^{-1} \tag{7.20}$$

The model is therefore a linear combination of the data. Data errors map into the model parameter estimates according to equation (6.51) with H replacing G and C_e

replacing C. The covariance of mapped errors is given by the rather complicated expression

$$\mathsf{HC_eH^T} = (\mathsf{A^TC_e^{-1}A} + \theta^2\mathsf{W})^{-1}\mathsf{A^TC_e^{-1}A}(\mathsf{A^TC_e^{-1}A} + \theta^2\mathsf{W})^{-1}, \tag{7.21}$$

but this is not the whole story. Equation (7.21) gives only that part of the model error mapped from the data errors. It takes no account of the uncertainty in the model arising from indeterminacy of the original equations. The problem is highlighted by the simple extreme case when the data contribute nothing to the solution. Then effectively $\mathsf{A} = 0$ and (7.21) gives zero. In other words, when there are no data the model estimate is exact! Furthermore, even when the data do define some model parameters well, (7.21) gives small errors for other, less well-determined parameters. Equation (7.21) must always give this: each model parameter is a linear combination of the data, and if that combination yields zero then the error must also map to zero.

A realistic estimate of model error should contain our *a priori* information, the knowledge that is independent of $\boldsymbol{d}$ and is embodied in the weight matrix W. Franklin (1970) developed a method he called *stochastic inversion*, in which the model parameters themselves are random variables with given prior covariance matrix $\mathsf{C_m}$. Identifying this with the inverse weighting:

$$\mathsf{C_m^{-1}} = \theta^2\mathsf{W}. \tag{7.22}$$

This means we are totally committed to precise weighting of the model: no variable trade-off parameter is allowed. We define a likelihood function

$$L = \frac{1}{2}\left[(\mathsf{A}\boldsymbol{m} - \boldsymbol{d})^{\mathrm{T}}\mathsf{C_e^{-1}}(\mathsf{A}\boldsymbol{m} - \boldsymbol{d}) + \boldsymbol{m}^{\mathrm{T}}\mathsf{C_m^{-1}}\boldsymbol{m}\right]. \tag{7.23}$$

Note first that setting $\partial L/\partial \boldsymbol{m} = 0$ gives the normal equations (7.16) with $\mathsf{C_m^{-1}}$ replacing $\theta^2\mathsf{W}$. Differentiating again gives

$$\frac{\partial^2 L}{\partial m_i \partial m_j} = \mathsf{A^TC_e^{-1}A} + \mathsf{C_m^{-1}}. \tag{7.24}$$

Equating this with C^{-1}, as we did in (6.48), gives the correct covariance matrix for the model parameters:

$$\mathsf{C} = (\mathsf{A^TC_e^{-1}A} + \mathsf{C_m^{-1}})^{-1}, \tag{7.25}$$

a much simpler form than (7.21).

In the undamped case we minimised the quadratic form $\boldsymbol{m}^{\mathrm{T}}(\mathsf{A^TC_e^{-1}A})\boldsymbol{m}$ with resulting model covariance $(\mathsf{A^TC_e^{-1}A})^{-1}$. In the damped case we minimise

$$\boldsymbol{m}^{\mathrm{T}}(\mathsf{A^TC_e^{-1}A} + \mathsf{C_m^{-1}})\boldsymbol{m}$$

with the resulting model covariance matrix C given by (7.25). When the data contain no information, C_{e} is very large and C reduces to C_{m}, the prior covariance estimate, as it should. Equation (7.25) should always be used, not the more complicated expression in (7.21). A rigorous treatment of the covariance matrix requires Bayes' theorem. See Box and Tiao (1973) or Bard (1974).

Yet another perspective on damping was found by Jackson (1979), who showed that adding new artificial data in terms of direct estimates of the model parameters themselves, with associated weights, gives exactly the same equations as damping. The new 'data' have the form $m_1 = a_1, m_2 = a_2, m_3 = a_3, \ldots$, with matrix form $\boldsymbol{a} = \mathsf{I}\boldsymbol{m}$, where I is the identity matrix. The data are given weight matrix $\mathsf{C}_{\mathrm{a}}^{-1}$. These *a priori* data augment the equations of condition matrix and change the A-matrix to the partitioned form

$$\mathsf{A} = \begin{pmatrix} \mathsf{A} \\ \hline \mathsf{I} \end{pmatrix}.$$

The normal equations matrix $\mathsf{A}^{\mathrm{T}}\mathsf{C}_{\mathrm{e}}{}^{-1}\mathsf{A}$ is replaced by $\mathsf{A}^{\mathrm{T}}\mathsf{C}_{\mathrm{e}}{}^{-1}\mathsf{A} + \mathsf{C}_{\mathrm{a}}^{-1}$, which has the same form as the damped normal equations (7.16) with weight matrix W replaced by $\mathsf{C}_{\mathrm{a}}^{-1}$. The right-hand sides of the normal equations are also changed. Jackson's method effectively constrains the solution for the model parameter to be close to the given model vector $\boldsymbol{a}$ and within the bounds allowed by the covariance matrix C_{a}.

Returning to problem (5), we now use some prior information to produce a core density larger than the mantle density. We could go further and use prior information that the core is made of liquid iron, which laboratory experiment shows to have a density of about 10^4 kg m^{-3}. It is possible to build inequality constraints into an inversion but it is tricky and awkward to compute, mainly because the inequality is nonlinear. A simpler approach is to redefine the norm to be minimised. Replacing $|\boldsymbol{m}|$ with $\sqrt{\rho_{\mathrm{c}}^2 + f\rho_{\mathrm{m}}^2}$, where $f > 1$ is chosen to produce the desired result. The larger f, the smaller the ratio $\rho_{\mathrm{m}}/\rho_{\mathrm{c}}$. The weight matrix W has the form

$$\mathsf{W} = \begin{pmatrix} 1 & 0 \\ 0 & f \end{pmatrix}.$$

A little experimentation shows that $f = 16$ gives a minimum norm solution of $(\rho_{\mathrm{c}}, \rho_{\mathrm{m}}) = (12\,283, 4088)$. Damping with $\theta = 0.02$ gives a solution $(\rho_{\mathrm{c}}, \rho_{\mathrm{m}}) = (9974, 3185)$, which is a plausible approximation to the Earth's density. The misfit is 1.32×10^{24} kg, much larger than the error estimate on the Earth's mass. This is, of course, because of the inadequacy of the simple model.

A device related to damping is the *penalty method*. We often need to apply some hard constraints to the model, for example to fix the depth of an earthquake because it is known to lie on a certain fault, freeze the magnetic field to the core fluid (an integral constraint), or fix the density of a particular rock unit. These constraints are normally imposed by the method of Lagrange multipliers (Appendix 6). A more convenient numerical approach is to use damping but make the damping constant so large that the model parameter (e.g. the earthquake depth) or constraint (combination of model parameters) is reduced effectively to zero. For example, in earthquake location the normal equations matrix is 4×4. Suppose the depth is component number 4 of the model vector. If at each iteration we add a big number, say 10^7 times any other matrix element, to the diagonal element of the fourth row of the normal-equations matrix, that equation effectively becomes $10^7 h = 0$ because the original elements of $\mathsf{A}^T\mathsf{A}$ and $\mathsf{A}^T\boldsymbol{d}$ are so much smaller than 10^7 they can be ignored. Thus at each iteration we have $h = 0$ and the depth never changes from its original value.

The penalty method (see also Appendix 6) is very easy to implement and has the advantage over Lagrange multipliers that you do not have to derive any new formulae, just add a big number to the right diagonal element. There is only one problem: deciding on an appropriate value of the damping constant or *penalty parameter*. It obviously depends on the problem and the units used to measure the model parameter, and must be found by trial and error. If the penalty parameter is too large the matrix becomes unbalanced and numerical errors result when the normal equations are solved; the effects that ensue from numerical error are often very peculiar and misleading.

Damping can also be used to great effect to constrain a general model that is already known. The following method was devised by Christoffersson and Husebye (1979), who wanted to impose some constraints on an inversion of seismic travel times using the known geology. Specifically, they wanted to equate all the seismic velocities within each of a few tectonic blocks within the study area. They called it their 'geological modelling' option. Their formalism is rather general and can be used to incorporate constraints into the normal equations after they have been formed. Since in most inverse problems with $D \gg P$ the time-consuming part of the computation is in doing the multiplication to form $\mathsf{A}^T\mathsf{A}$, rather than in finding A or solving the normal equations[1], the method allows a large number of different ideas to be tested with a single computation. In their example, a tomographic inversion of NORSAR seismic array data, they first parametrise the seismic velocity beneath the array and find the normal equations. Their constraints are then to equalise all the velocities in individual tectonic zones, effectively reducing the number of model parameters and testing the idea that seismic velocity can be deduced from the geology.

[1] Actually forming the normal equations in this way is not recommended computational practice because it squares the condition number of the matrix; the approved method is to solve the equations of condition directly by rotational methods. The method of Christoffersson and Husebye can be adapted to work equally well with either numerical method.

The original model vector $\boldsymbol{m}$ is reduced to a smaller vector $\boldsymbol{p}$ through the constraints equation

$$\boldsymbol{m} = \mathsf{G}\boldsymbol{p}. \tag{7.26}$$

For example, to equalise two of the original model parameters and to zero a third, the equations would be:

$$\begin{pmatrix} m_1 \\ m_2 \\ m_3 \\ m_4 \end{pmatrix} = \begin{pmatrix} 1 & 0 \\ 1 & 0 \\ 0 & 0 \\ 0 & 1 \end{pmatrix} \begin{pmatrix} p_1 \\ p_2 \end{pmatrix}. \tag{7.27}$$

Christoffersson and Husebye (1979) then show that the (unweighted) normal equations become

$$\mathsf{G}^{\mathrm{T}}(\mathsf{A}^{\mathrm{T}}\mathsf{A})\mathsf{G}\boldsymbol{p} = \mathsf{G}^{\mathrm{T}}\mathsf{A}^{\mathrm{T}}\boldsymbol{d}, \tag{7.28}$$

which may be solved for the reduced model vector $\boldsymbol{p}$. The weighted, reduced normal equations can easily be derived in the same way.

7.6 The resolution matrix

Resolution is the most important concept distinguishing inverse theory from parameter estimation. The damped solution satisfies equation (7.16) whereas the undamped ('true') solution must satisfy (6.59). Denoting the undamped solution in this case by $\boldsymbol{m}_{\mathrm{t}}$, combining (6.59) and (7.16), and replacing $\theta^2\mathsf{W}$ with $\mathsf{C}_{\mathrm{m}}^{-1}$ gives

$$\boldsymbol{m} = \mathsf{R}\boldsymbol{m}_{\mathrm{t}} \tag{7.29}$$
$$\mathsf{R} = (\mathsf{A}^{\mathrm{T}}\mathsf{C}_{\mathrm{e}}^{-1}\mathsf{A} + \mathsf{C}_{\mathrm{m}}^{-1}(\mathsf{A}^{\mathrm{T}}\mathsf{C}_{\mathrm{e}}^{-1}\mathsf{A})), \tag{7.30}$$

where R is a $P \times P$ square, non-symmetric, array called the *resolution matrix*. Note that the result is not symmetric because the product of two symmetric matrices is only symmetric if they commute.

The resolution matrix relates the estimate to the true model; note that it cannot be inverted (otherwise we would be able to find the true model and all our problems would be over!). Resolution is perfect when $\mathsf{R} = \mathsf{I}$ and there is no difference between the true model and the estimate. This happens when $\mathsf{C}_{\mathrm{m}}^{-1} = 0$.

Consider the ith equation of (7.29). Written out in full it gives

$$m_i = \sum_{j=1}^{P} R_{ij}(\boldsymbol{m}_{\mathrm{t}})_j. \tag{7.31}$$

The estimate of each component m_i is therefore a weighted average of the entire true model. The ith row of R gives the weights for the ith model parameter m_i. Ideally we want $R_{ii} = 1$ and all other weights zero, which is the case when $\mathsf{R} = \mathsf{I}$. The ith row of the resolution matrix is the *averaging function* for the ith model parameter. Each model parameter has a different averaging function given by the corresponding row of the resolution matrix.

Non-zero off-diagonal elements of the resolution matrix indicate a trade-off between the model parameters: non-zero R_{ij} indicates failure to separate the effect of the ith and jth model parameters. Note that R does not depend directly on data errors (except in so far as C_e ensures that they are correctly weighted relative to each other, and in the choice of trade-off parameter θ), but it does depend on A through the data kernels. Effectively this means that resolution depends on where the data were measured rather than on how well they were measured. In a seismic experiment, for example, the elements of R will depend explicitly on the geographical positions of the seismometers but not on the noise level in the seismograms. Resolution is generally determined by the *type* of data collected, not by their accuracy. Resolution is therefore an important consideration when designing an experiment.

The trace of the resolution matrix (sum of its diagonal elements = sum of its eigenvalues) is often taken to represent the *number of degrees of freedom* in the model. Ideally it is P when the resolution is unity; otherwise it is less than P. The number of degrees of freedom is also thought of as the number of parameters that can be resolved by the dataset. If the diagonal elements of the resolution-matrix elements were entirely ones or zeroes the trace would correspond exactly to the number of parameters, but in general each parameter is only partially resolved.

There is an efficient way to compute the resolution matrix in the usual case when $\mathsf{C}_\mathrm{m}^{-1}$ is diagonal. First note from (7.30) and (7.25) that the resolution matrix is simply the covariance matrix times the normal-equations matrix:

$$\mathsf{R} = \mathsf{C}\mathsf{A}^\mathrm{T}\mathsf{C}_\mathrm{e}^{-1}\mathsf{A}. \tag{7.32}$$

If $\mathsf{C}_\mathrm{m}^{-1}$ is diagonal a much faster method to evaluate R is to rewrite (7.30) in the form

$$\begin{aligned} \mathsf{R} &= (\mathsf{A}^\mathrm{T}\mathsf{C}_\mathrm{e}^{-1}\mathsf{A} + \mathsf{C}_\mathrm{m}^{-1})^{-1}(\mathsf{A}^\mathrm{T}\mathsf{C}_\mathrm{e}^{-1}\mathsf{A} + \mathsf{C}_\mathrm{m}^{-1} - \mathsf{C}_\mathrm{m}^{-1}), \\ \mathsf{R} &= \mathsf{I} - \mathsf{C}\mathsf{C}_\mathrm{m}^{-1}. \end{aligned} \tag{7.33}$$

The final matrix multiplication of a symmetric matrix with a diagonal matrix is very fast. The recipe for the under-determined inverse problem is given in Box 7.1.

Box 7.1. Summary: recipe for the under-determined inverse problem

The following procedures are in addition to those listed for parameter estimation in Box 6.1.

(i) Establish the extent of the null space, and preferably its dimension. This may be possible from the nature of the original problem, such as $N = P - D$ when there are no complications, but this is usually not possible in practice.
(ii) Examine the eigenvalues of the normal-equations matrix A. If there is a clear null space some eigenvalues of the spectrum will be numerically zero. Unfortunately, for large problems, there is usually a large set of small eigenvalues that must be stabilised, many of which are numerically zero.
(iii) If there is an obvious break in the spectrum of eigenvalues, with a clearly distinct set of zeroes, then winnow by removing them. Otherwise it is usually easier to leave them in and rely on simple damping to remove their effect on the solution.
(iv) Decide on the nature of the damping. Use only the physical properties of the problem, not information derived from inadequacy of the data at hand. Ideally construct $\mathsf{C}_{\mathrm{m}}^{-1}$ from prior information.
(v) If necessary introduce a damping constant. Make sure all model parameters are of equal dimension and scale, or choose different damping constants accordingly.
(vi) Invert for the model parameters using equation (7.16).
(vii) Compute the norm for each value of the damping constant and plot a trade-off curve to establish consistency of the prior information with the error estimates and model itself.
(viii) Compute the resolution matrix and its trace.

EXERCISES

7.1. Find the eigenvalues and normalised eigenvectors of the matrix

$$\mathsf{G} = \begin{pmatrix} 1 & 0 & 1 \\ 0 & 1 & 0 \\ 1 & 0 & 1 \end{pmatrix}.$$

What is the null space? Plot it on a diagram.

7.2. Find the null space of the matrix

$$\begin{pmatrix} 1 & 1 & 1 \\ 1 & 1 & 1 \\ 1 & 1 & 1 \end{pmatrix}.$$

7.3. Find, using the eigenvector expansion, the general solution of the equation $\mathsf{A}\boldsymbol{x} = \boldsymbol{b}$, where $\boldsymbol{b} = (6, 6, 6)^{\mathrm{T}}$ and A is the matrix in the previous question. Obtain the minimum-norm solution. Why do equations (7.6) and (7.7) not work in this case?

7.4. By rewriting the equations of the previous question as $(1, 1, 1)\boldsymbol{x} = 6$, find the minimum-norm solution using (7.7) and compare it with the answer of the previous question. Verify that your generalised inverse satisfies the appropriate form of (7.8).

7.5. Given the simultaneous equations:

$$200x + 4y - 2z = 202$$
$$4x + 4y - 2z = 6$$
$$-2x - 2y + z = -3$$

find, by inspection or otherwise, the combination of parameters that cannot be determined. Using `octave`, or otherwise, rank the eigenvalues and find the corresponding eigenvectors. Obtain the solutions when one, then two, eigenvalues are winnowed out.

7.6. **The remaining questions in this chapter are computer exercises.**
Consider the previous question as a linear inverse problem. Only the first two equations are used: there are two measurements (202 and 6) and three unknowns. $\boldsymbol{x}$ is the model and $\boldsymbol{d}$ is the data vector. Each measurement is assigned an error of 1.0. Solve for the model by simple damping.

(a) Use `octave` to compute the solution, model norm, and rms misfit for a range of damping parameters.
(b) Plot the trade-off curve.
(c) Find the value of the damping parameter giving unit weighted misfit and compare the corresponding solution with those of the previous question.
(d) Calculate the parameter covariance matrix for this value of the damping parameter.
(e) Calculate the resolution matrix for the same solution. How well resolved are each of the unknowns?

7.7. Fit a cubic to the following data by ordinary least squares parameter estimation: $(x, y) = (0, 3), (0.5, 0), (1, -0.5), (2, 1.2), (3, 0)$. Errors are all 0.5. Calculate the rms misfit and model norm. Do the data justify this cubic fit, or could a smoother curve do an adequate job?

You have a rough theory that predicts the plot should be a straight line. Damp the solution to make it smoother with a damping matrix that penalises the higher powers of x the most. Use a diagonal damping matrix with elements $(0, 0, 4, 16)$ (why?). Find the value of the damping constant that gives unit misfit to the weighted data, and compare the corresponding solution to the ordinary least-squares solution.

8

Nonlinear inverse problems

8.1 Methods available for nonlinear problems

So far we have restricted ourselves to problems where the relationship between the data and the model is linear. The vast majority of forward problems are not linear: these are lumped together and described as 'nonlinear', an accurate but not very useful classification because there is no general method for solving nonlinear inverse problems. Available methods are divided here into three broad classes: *tricks* that work for individual problems; *forward modelling*, or brute force solution of the forward problem for many candidate model parameters; and *linearisation*, or approximating the nonlinear forward problem with a linear equation and the use of linear methods.

This is a good time to remind ourselves of the difference between simply finding a model that explains the data, and solving the entire inverse problem. Solving the entire inverse problem involves finding the whole set of models that can explain the data. Finding the model is an exercise in numerical analysis, a vast subject well treated elsewhere, for example in the excellent *Numerical Recipes* (Press *et al.*, 1992). Solving the inverse problem means finding all possible models that fit the data, a much larger task.

Little can be said in general about special 'tricks' that work for just one problem, save to mention that the commonest device is to notice that an apparently nonlinear problem is really a disguised linear problem. A recent example is determination of the geomagnetic field from purely directional data or from intensities alone (Khokhlov *et al.*, 2001).

An all-out attack using forward modelling starts with an efficient algorithm for solving the nonlinear forward problem, effectively predicting the data from a chosen set of model parameters. This allows us to compute the misfit (for example the sum of squares of errors) in order to assess the quality of fit and suitability of the chosen model. We then choose another set of model parameters and determine its quality of fit to the data, repeating the procedure until we deem a sufficient number of models has been searched. This approach may not be elegant but it gives a great deal of information.

Forward modelling might seem simple given a large computer, but the workload can rapidly become impractically large without a good strategy for choosing suitable models. Even a model with a single parameter needs an infinite number of forward calculations to define the full solution of the inverse problem. Except in a tiny minority of cases where the forward problem has an analytical solution, it is only possible to examine a finite number of models. Two decisions must be made: how finely should the model space be sampled, and where in that model space should we concentrate?

The strategy depends on the problem at hand but there are two distinct approaches. A *grid search* is a systematic exploration of model space with a given increment in model parameters, starting from some centre and working out until a sufficient volume of the space has been searched, perhaps when the misfit exceeds some chosen value or the model is deemed unphysical for other reasons. An example of a grid search is given in the beamforming application in Chapter 11, where a seismogram is stepped successively forwards and backwards in time until it matches another seismogram.

Monte Carlo methods use some random component of selection. The choice can be guided by prior information about the likelihood of a particular model fitting the data, so the search can be concentrated in certain parts of model space yet still retain some random element of choice. These methods were pioneered for seismic inversions for earth structure. If a prior probability distribution can be attached to a particular choice of model the procedure can be put on a firm statistical footing. This approach is treated by Tarantola (1987). A number of modern algorithms employ more sophisticated strategies for searching model space. Most notable is the suite of genetic algorithms (see, for example, Sambridge and Gallagher (1993)), in which candidate models are selected according to a set of rules relating them to successful models computed previously, the rules being akin to genetic inheritance.

Linearisation is still probably the most-used method for solving nonlinear inverse problems. We first choose a likely starting model, compute the residuals to the data, then seek a small change in the model that reduces the residuals. The new model can be used to form the basis of another small improvement, and the procedure repeated until a model is found that fits the data. This completes one part of the inversion process, finding a model. A simple example of this procedure is Newton–Raphson iteration for a root of a nonlinear algebraic equation.

Finding the model in this way should strictly be called *quasi-linearisation* because we arrive at the solution to the nonlinear problem without incurring any errors caused by the linearisation process. Solving the inverse problem, however, involves mapping out all the possible model solutions and does depend on the linearising approximation; it requires that errors on the data are small in some sense. Provided the errors on the data are sufficiently small, the linear error analysis for our final, best-fitting model will apply to the nonlinear problem. We can then employ the theory developed in Chapters 6 and 7 to produce estimates of covariance and resolution. If the errors are too large for the linear approximation to be valid there is rather little we can do. A correct approach is to consider a higher approximation, adding quadratic terms to the linear approximation for example, as Sabatier (1979) has done. We then need also to consider higher statistical moments than the covariance, but these are rarely available. It is better to return to forward modelling.

There remains another possibility: that completely different models satisfy the data. These will be surrounded by their own subspaces of acceptable models. These model subspaces may or may not overlap. Note once again that the conditions required for convergence of the quasi-linear iterations are completely different from those required for the validity of the linearised error analysis. One depends only on the starting model and the nature of the forward problem, the other depends on the errors in the data. If the errors are too large, in the sense that the linearising approximation ceases to be valid, the model covariance matrix formula (7.25) fails and higher-order moments are needed to describe the true error.

8.2 Earthquake location: an example of nonlinear parameter estimation

Earthquake location involves finding the origin time, depth, latitude and longitude of an assumed point source of seismic waves, the hypocentre. It is an example of parameter estimation (Chapter 6) with four parameters. Arrival times of P and S waves are measured from stations with a broad geographical distribution and inverted for the hypocentre parameters. The forward problem involves calculating travel times of the waves, which requires a known velocity structure or set of travel-time tables. The problem is nonlinear because the travel times do not depend linearly on the end points of the ray, the hypocentre coordinates.

We begin with an approximate location and origin time. We can trace the ray from this source to the receiver by using the seismic velocity as a function of position, compute the travel times, and add the origin time to give the predicted arrival times. Alternatively, given a set of travel-time tables, we can look up the travel time for the appropriate depth and distance. Subtracting the computed arrival time from the observed time gives the residual we wish to reduce by improving the location.

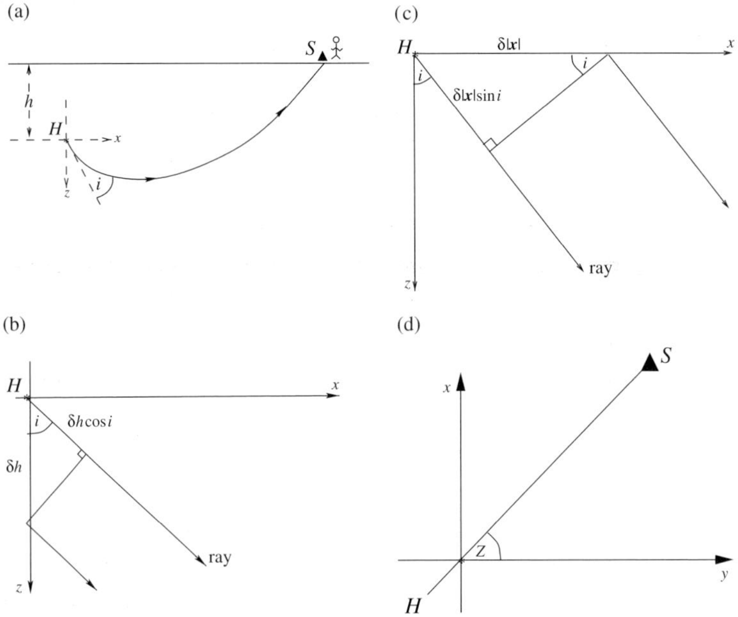

Fig. 8.1. Ray geometry showing how to compute the partial derivatives for hypocentre improvement. (a) Hypocentral coordinates; (b) ray shortening with changes in depth; (c) ray shortening with changes in epicentral location; and (d) plan view for resolving epicentral movement along x- and y-directions.

We now calculate the change in arrival time caused by a small change in origin time and hypocentre location using the ray geometry illustrated in Figure 8.1(a). We require the take-off angle of the ray, i, which may be computed from the tangent to the ray path at the hypocentre or, in the usual case when the seismic velocity depends only on depth, from the travel-time tables using standard methods described in, for example, Bullen and Bolt (1985).

Changing the origin time T_O by δT_O simply changes the arrival time by δT_O. This part of the problem is linear: the change in arrival time is linearly related to the origin time. Changing the depth h by a small amount δh shortens the ray path by $\delta h \cos i$, from the triangle in Figure 8.1(b), which shortens the travel time by $\delta h \cos i / v_H$, where v_H is the seismic velocity at the hypocentre. The linearised change in arrival time corresponding to unit change in hypocentre depth is therefore $-\cos i / v_H$. Changing the epicentre (x, y), where axes x, y may be taken, for example, North and East with origin at the epicentre, by small amount $(\delta x, \delta y) = \delta \boldsymbol{x}$ shortens the ray path by $|\delta \boldsymbol{x}| \sin i$ and the arrival time by $|\delta \boldsymbol{x}| \sin i / v_H$ (Figure 8.1(c)). The ray path lies in the plane of the source and the receiver, making an angle Z with the x-axis (Figure 8.1(d)). Z is the azimuth of the receiver from the source, which may be computed from the geometry. The reduction in arrival time caused by a small change δx in epicentre is then $\sin i \cos Z \delta x / v_H$ and for δy

is $\sin i \sin Z\delta y/v_H$. The changes in arrival time corresponding to unit change in x and y are then $-\sin i \cos Z/v_H$ and $-\sin i \sin Z/v_H$ respectively.

We have effectively calculated the *partial derivatives* of the travel time with respect to each of the four hypocentral coordinates. In the next section we write down a general form for the forward problem and compute these partial derivatives by differentiation. For earthquake location the general form of the travel times is a complicated integral and it is easier to work out the partial derivatives graphically as we have done: the linear approximation amounts to assuming the ray path is a straight line in the vicinity of the hypocentre.

The partial derivatives are functions of the ray path and are therefore different for each datum. The predicted change in arrival time at station 1 is then

$$\delta T_1 = \delta T_O - \frac{\cos i_1}{v_H}\delta h - \frac{\sin i_1 \cos Z_1}{v_H}\delta x - \frac{\sin i_1 \sin Z_1}{v_H}\delta y. \tag{8.1}$$

We now form the standard equation for a linear inverse problem (6.4) by placing the arrival time residuals into a data vector $\delta \boldsymbol{d}$, the small changes to the hypocentre into a model vector $\delta \boldsymbol{m} = (\delta T_O, \delta h, \delta x, \delta y)$, and the partial derivatives into the equations of condition matrix A:

$$\delta \boldsymbol{d} = \mathsf{A}\delta \boldsymbol{m} + \boldsymbol{e}. \tag{8.2}$$

The nth row of this matrix is then

$$A_{n*} = \left(1, -\frac{\cos i_n}{v_H}, -\frac{\sin i_n \cos Z_n}{v_H}, -\frac{\sin i_n \sin Z_n}{v_H}\right). \tag{8.3}$$

Subscript n corresponds to a particular station and type of arrival (P or S, for example). It labels the ray path, which is different for each station and, in general, for each wave type. A is a *matrix of partial derivatives*; it is rectangular of size $D \times 4$, where D is the number of data. The first column of A consists entirely of ones.

The standard least-squares solution is found as described for linear problems in Chapter 6. The resulting small changes to the hypocentre are added to the original estimate to give the new estimate, which in turn can form the starting point for a second iteration. The partial derivatives must be recomputed because the ray path has changed; i, Z, and v_H will all be different. The process is deemed to have converged when the hypocentre estimate ceases to change significantly. Deciding on what 'significantly' means may prove problematic: a logical choice would be when the change in residuals falls significantly below their estimated error. The last step in the process may be used to derive the covariance matrix for the small change $\delta \boldsymbol{m}$. If the data errors lie within the linear range the linearised covariance matrix gives an accurate error estimate for the final hypocentre.

It often happens that equations (8.2) are ill-conditioned, prohibiting a numerical solution. For example, the data may not constrain the depth, or there may be a trade-off between depth and origin time because of an insufficient density of stations close to the earthquake. The iteration procedure fails to converge. In these cases damping techniques similar to those described in Chapter 7 may be used to stabilise the solution. We must be careful to distinguish this type of *step-length damping*, which is used to make the iterative procedure converge, from damping used to control the final fit to the data. The two forms of damping serve fundamentally different purposes, despite sharing the same mathematical formulation. Step-length damping is essentially a numerical device to aid convergence; it may have nothing to do with data or model error.

8.3 Quasi-linearisation and iteration for the general problem

Write the general relationship between data and model as

$$\boldsymbol{d} = \boldsymbol{F}(\boldsymbol{m}) + \boldsymbol{e}, \tag{8.4}$$

where $\boldsymbol{F}$ denotes a nonlinear function or formula. For the example of earthquake location, $\boldsymbol{F}$ represents the whole process of ray tracing or table look-up to determine the travel time followed by addition to the origin time to give an arrival time at each station. We initiate the iteration from some starting model $\boldsymbol{m}_0$ and seek a small improvement $\delta \boldsymbol{m}_0$ by linearised inversion of the residuals $\delta \boldsymbol{d}_0 = \boldsymbol{d} - \boldsymbol{F}(\boldsymbol{m}_0)$:

$$\delta \boldsymbol{d}_0 = \mathsf{A}_0 \delta \boldsymbol{m}_0 + \boldsymbol{e}. \tag{8.5}$$

The matrix A_0 must be derived from the nonlinear formula $\boldsymbol{F}(\boldsymbol{m})$ for each iteration; subscript zero denotes the first iteration. The elements of A are partial derivatives of the data with respect to the model, and may be written formally as

$$(\mathsf{A}_0)_{ij} = \left. \frac{\partial F_i}{\partial m_j} \right|_{\boldsymbol{m}=\boldsymbol{m}_0}. \tag{8.6}$$

In the earthquake location example it was easier to derive the elements of A from first principles using the geometry of the ray path. It is usually more straightforward to write down the forward problem and carry out the necessary differentiation of $\boldsymbol{F}$. For the earthquake location problem $\boldsymbol{F}$ is a path integral and we would have to differentiate with respect to the end point, the lower limit of the integral.

The linearised equations of condition (8.2) are solved for $\delta \boldsymbol{m}_0$, by using (6.25) or (6.59) in the same way as for the linear inverse problem. The improved solution is

$$\boldsymbol{m}_1 = \boldsymbol{m}_0 + \delta \boldsymbol{m}_0. \tag{8.7}$$

If $\boldsymbol{m}_1$ does not provide a good enough solution the whole procedure is repeated, starting with $\boldsymbol{m}_1$ instead of $\boldsymbol{m}_0$. The general equations for the Ith iteration are

$$\delta \boldsymbol{m}_I = \left(\mathsf{A}_I^{\mathrm{T}} \mathsf{C}_{\mathrm{e}}^{-1} \mathsf{A}_I\right)^{-1} \mathsf{A}_I^{\mathrm{T}} \mathsf{C}_{\mathrm{e}}^{-1} [\boldsymbol{d} - \boldsymbol{F}(\boldsymbol{m}_I)] \tag{8.8}$$

$$\boldsymbol{m}_{I+1} = \boldsymbol{m}_I + \delta \boldsymbol{m}_I. \tag{8.9}$$

In general the iteration procedure is repeated until $|\boldsymbol{m}_{I+1} - \boldsymbol{m}_I|$ is less than some prescribed value, when further iterations fail to produce significant change in the model. An alternative convergence criterion is to require $|\boldsymbol{F}(\boldsymbol{m}_{I+1}) - \boldsymbol{F}(\boldsymbol{m}_I)|$ to be less than some prescribed value. This is the improvement in fit to the data achieved by the Ith iteration. It is better than using convergence of the model for two reasons. First, we have a clear idea of how small this number should be from our estimates of the errors in the data. Secondly, there is little point in continuing to iterate when there is no further improvement in fit to the data even though the model is still changing: further changes have no useful effect.

8.4 Damping, step-length damping, and covariance and resolution matrices

Having found the best-fitting model, the covariance and resolution matrices can be found in the same way as for linear inversion by using equations (7.25) and (7.30). Essentially we perform a linear inversion about the final model $\boldsymbol{m}$ and discard the model change $\delta \boldsymbol{m} \approx 0$, and calculate the uncertainties on $\delta \boldsymbol{m}$. The procedure will only be valid provided the allowed space of models defined by the covariance matrix lies within the linear regime, in other words the linear approximation (8.2) is accurate in predicting changes to the data for changes to the model within its uncertainty.

In a linear problem damping determines the model that minimises the combination of error and model norm:

$$T = \boldsymbol{e}^{\mathrm{T}} \mathsf{C}_{\mathrm{e}}^{-1} \boldsymbol{e} + \theta^2 \boldsymbol{m}^{\mathrm{T}} \mathsf{W} \boldsymbol{m} \tag{8.10}$$

(cf. equations (7.15) and (7.16)). If we simply apply the linear formulae to each nonlinear iteration we would damp not the model itself, $\boldsymbol{m}$, but the small change imposed at each iteration, $\delta \boldsymbol{m}$ (step-length damping). To solve the inverse problem posed by minimising (8.10) we must use (8.5) for $\boldsymbol{e}$ and $\boldsymbol{m}_0 + \delta \boldsymbol{m}$ for $\boldsymbol{m}$. The basic difference with the linear problem is illustrated by the simple case of unweighted data and unweighted model parameters. We minimise

$$T = \boldsymbol{e}^{\mathrm{T}} \boldsymbol{e} + \theta^2 \boldsymbol{m}^{\mathrm{T}} \boldsymbol{m} \tag{8.11}$$

$$= \sum_{i=1}^{D} \left(\delta d_i - \sum_{j=1}^{P} A_{ij} \delta m_j \right)^2 + \theta^2 \sum_{i=1}^{D} (m_{0i} + \delta m_i)^2 \tag{8.12}$$

by differentiating T with respect to the unknowns $\delta \boldsymbol{m}$ to give

$$\frac{\partial T}{\partial \delta m_n} = 2\sum_i \left(\delta d_i - \sum_j A_{ij}\delta m_j \right)(-A_{in}) + \theta^2(m_{0n} + \delta m_n) = 0. \quad (8.13)$$

In matrix form this equation rearranges to

$$-(\mathsf{A}^{\mathrm{T}}\delta \boldsymbol{d} - \mathsf{A}^{\mathrm{T}}\mathsf{A}\delta \boldsymbol{m}) + \theta^2(\boldsymbol{m}_0 + \delta \boldsymbol{m}) = 0 \quad (8.14)$$

and therefore $\delta \boldsymbol{m}$ satisfies

$$(\mathsf{A}^{\mathrm{T}}\mathsf{A} + \theta^2\mathsf{I})\delta \boldsymbol{m} = \mathsf{A}^{\mathrm{T}}\delta \boldsymbol{d} - \theta^2 \boldsymbol{m}_0 \quad (8.15)$$

and the solution for the model after this iteration is

$$\begin{aligned} \boldsymbol{m}_1 &= \boldsymbol{m}_0 + \delta \boldsymbol{m} \\ &= \boldsymbol{m}_0 + (\mathsf{A}^{\mathrm{T}}\mathsf{A} + \theta^2\mathsf{I})^{-1}(\mathsf{A}^{\mathrm{T}}\delta \boldsymbol{d} - \theta^2 \boldsymbol{m}_0). \end{aligned} \quad (8.16)$$

The last term on the right-hand side is new; it must be included in order to damp the final model rather than each linear step. Minimising (8.10) gives

$$\delta \boldsymbol{m} = \left(\mathsf{A}^{\mathrm{T}}\mathsf{C}_{\mathrm{e}}^{-1}\mathsf{A} + \theta^2\mathsf{W}\right)^{-1}\left(\mathsf{A}^{\mathrm{T}}\mathsf{C}_{\mathrm{e}}^{-1}\delta \boldsymbol{d} - \theta^2\mathsf{W}\boldsymbol{m}_0\right). \quad (8.17)$$

Comparing with the solution for the linear problem (7.16) shows the additional term $-\theta^2\mathsf{W}\boldsymbol{m}_0$. If a model covariance matrix is used, $\theta^2\mathsf{W}$ should be replaced by $\mathsf{C}_{\mathrm{m}}^{-1}$ in these equations. The covariance and resolution matrices need no extra terms in the nonlinear case because they assume a one-step linear inversion from the final model: equations (7.25) and (7.30) stand.

8.5 The error surface

A useful geometrical analogue is the *error surface*, the error plotted as a function of the model parameters:

$$E(\boldsymbol{m}) = [\boldsymbol{d} - \boldsymbol{F}(\boldsymbol{m}]^{\mathrm{T}}[\boldsymbol{d} - \boldsymbol{F}(\boldsymbol{m})]. \quad (8.18)$$

The geometrical shape of the error surface tells us the nature of the minima we seek, and can explain the behaviour of the iterative approach to the best-fitting model.

The error 'surface' for a one-parameter model is a line. For a linear problem the error surface is a parabola (Figure 8.2(a)). The best-fitting model is at the minimum of the parabola, m_{f}. The data errors set an allowed misfit, at level A in the figure, which defines the range of allowed models, m_- to m_+.

For a nonlinear problem the error surface can have any shape, but we hope the problem is reasonably well posed so that the error surface has some well-defined

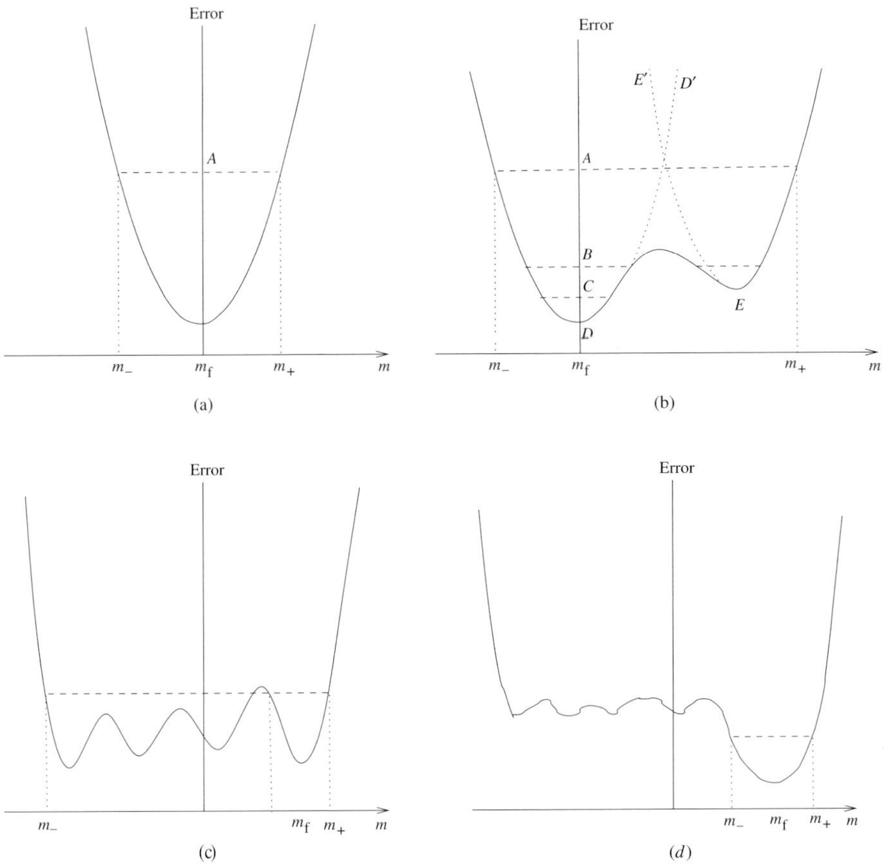

Fig. 8.2. Possible 1D error surfaces. (a) Quadratic, a linear problem; (b) two local minima; (c) many local minima; and (d) a relatively flat surface with one well-defined local minimum.

minima and the error increases strongly when the model parameters take on extreme values. Many possibilities exist. The most convenient is when the error surface has a single minimum with a quadratic-like shape, and the data errors confine the allowed solutions to be close to the bottom of the minimum. Under these conditions the linearisation described in the last section is valid; this is the scenario we hope for in most nonlinear examples. If the data are good enough, these conditions hold.

A more serious problem arises when there is more than one minimum to the error curve, as in Figure 8.2(b). The iterative procedure could end up in either of these minima; in particular we could have arrived in the shallower one without knowing about the existence of the deeper minimum to the left. The 'best-fitting' model is only a *local* best fit; the *global* best fit is the solution to the left. There are three possible outcomes to the error analysis, depending on the data errors. If the errors are small, at the level C in Figure 8.2(b), and we happen to have found the global minimum, then we have the right solution and an accurate estimate of the uncertainties. If the iterative procedure lands us in the shallower minimum we would conclude from the error analysis that the model was inconsistent with the data because the misfit is greater than C.

Suppose we have noisier data, level B in Figure 8.2(b). Then both minima are consistent with the data and we should accept all models lying within either potential well. The linearised error analysis in each potential well will be based on the parabolae ending in D' and E', which give reasonable approximations to the truth. With even noisier data at level A, the whole well will define acceptable model solutions. An error analysis based on one of the two local minima will be a serious underestimate because they will follow one of the two parabolae labelled D' or E'.

Figure 8.2(c) gives another possibility: multiple minima of equal depth. This situation arises frequently when trying to match oscillatory time series – two seismograms from the same source, for example. Each seismogram consists of a number of oscillations with the same frequency or similar shape but different phase. An idealisation of this problem is given by the two functions $f_1(t) = \cos \omega t$ and $f_2(t) = \cos(\omega t + \epsilon)$. We wish to find the time difference ϵ by minimising the error

$$E(\epsilon) = \int [f_1(t) - f_2(t + \epsilon)]^2 \mathrm{d}t, \qquad (8.19)$$

where the integral is taken over the length of available time series, probably many periods $2\pi/\omega$. The error has a minimum (zero in this case) at $\epsilon = 0$ and other subsidiary minima when ϵ is a multiple of the period $2\pi/\omega$. These subsidiary minima occur when the wrong peak in one record is lined up with one of the peaks in the first record. It is called *cycle skipping* and gives rise to error surfaces of the form shown in Figure 8.2(c).

Finally, a very flat error surface will mean the iterative procedure will converge very slowly. Figure 8.2(d) shows an example; in this case the inverse problem has a satisfactory solution provided we can find the smallest minimum. If the whole error surface is rather flat and there is no distinct minimum, the data errors will map into large errors in the model parameter. In this case we should probably spend our time making new measurements rather than trying to invert the ones we have!

The geometrical analogy stands up well in higher dimensions when we have more model parameters. To find the model we need to find a hollow in the error surface, and once there we approximate it with a multidimensional quadratic surface and use the data errors to determine the depth and acceptable space of models. Many iterative procedures have been devised to solve nonlinear problems;

the one described here is frequently used but is not always the best. Modern subspace methods, such as the conjugate-gradient algorithm, are more efficient for large problems. Again, this is the realm of numerical analysis rather than inverse theory. The recipe for the nonlinear inverse problem is given in Box 8.1.

Box 8.1. Summary: recipe for the nonlinear inverse problem

The following procedures are in addition to those listed for parameter estimation in Box 6.1 and the under-determined inverse problem in Box 7.1.

(i) Is there an analytic solution? Is it a disguised linear solution?
(ii) Probably not. Decide on whether to use a quasi-linearised iteration or forward modelling. This depends on whether you think you have a good estimate of the final answer, in which case quasi-linearisation is best, and on the difficulty of solving the forward problem a very large number of times. You may decide to use a hybrid method (see Section 11.1), and start with forward modelling to obtain a starting model of quasi-linearisation.
(iii) If you decide on forward modelling, you need a strategy for choosing sample models. This depends critically on the specific problem. New methods, such as genetic algorithms, may be appropriate.
(iv) Linearise the problem by carrying out the necessary differentiation to obtain the matrix of partial derivatives.
(v) Employ step-length damping if a single iteration produces an unacceptably large change to the model. The change should typically be smaller than the original model parameters.
(vi) Damping for the final iteration determines the error analysis.
(vii) Determine the size of the model parameter error ellipsoid in relation to the size of the linear regime. If models within the space of allowed solutions are too far from the minimum norm solution for the linear approximation to predict the data accurately the error analysis is useless and you must revert to forward modelling.

EXERCISES

8.1. Investigate the effect of the starting guess on the simple problem of finding the roots of a quartic by Newton–Raphson iteration. First plot the quadratic

$$f(x) = 3x^4 - 16x^3 + 18x^2$$

and identify the approximate locations of the four real roots (solutions of $f(x) = 0$). Next solve for the roots and stationary points analytically.

Now solve for the roots using Newton–Raphson iteration with starting values ranging from -1 to 4. (You may find the FORTRAN program `newtfind.f` on the web site useful). Draw up a table with three columns, one containing the starting value, one the root found (if any), and one the number of iterations taken to reach four significant figures. Pay particular attention to the behaviour with starting points near the stationary values.

Explore ways to improve on the convergence of the difficult cases. Try step-length damping by reducing the change in x at each iteration. If this does not work well, try reducing the changes for the first few iterations and then proceed without damping.

8.2. How would you determine the parameter a when fitting data $\boldsymbol{d}$ with errors $\boldsymbol{e}$ to a model with parameter a given the following relationship?

$$d(x_i) = \exp a x_i$$

8.3. Solve the following nonlinear inverse problem using the formalism in Section 8.3:

$$\boldsymbol{d}_i = a^2 x_i^2 + a x_i + 1 + e_i,$$

where a is the (one-dimensional) model, $\boldsymbol{e}$ is the error vector, and $\boldsymbol{x}$ and the data vectors are given as follows.

x	0.0	0.1	0.2	0.3	0.4	0.5	0.6	0.7	0.8	0.9	1.0
d	1.01	1.22	1.57	2.01	2.39	2.97	3.71	4.33	5.20	6.00	7.02

8.4. For the previous question, find the variance of a if the errors are uncorrelated with variance 0.002. Plot the error surface around your solution. Does the variance estimate lie within the linear regime?

8.5. The first step in any inverse problem is to solve the forward problem. For LINEAR inverse problems this involves finding the form of the equations $\boldsymbol{d} = \mathsf{A}\boldsymbol{m}$. Find the row of the equations of condition matrix A corresponding to each of the following data for the given inverse problem.

(a) The geomagnetic field is represented by a model vector consisting of the Gauss or geomagnetic coefficients:

$$\boldsymbol{m}^T = (g_1^0, g_1^1, h_1^1, g_2^0, g_2^1, h_2^1, g_2^2, \ldots).$$

The geomagnetic potential is given by

$$V(r, \theta, \phi) = a \sum_{l,m} \left(\frac{a}{r}\right)^{l+1} \left(g_l^m \cos m\phi + h_l^m \sin m\phi P_l^m\right)(\cos\theta)$$

and the cartesian components by

$$X = -\frac{1}{r}\frac{\partial V}{\partial \theta}$$

$$Y = \frac{1}{r \sin\theta}\frac{\partial V}{\partial \phi}$$

$$Z = -\frac{\partial V}{\partial r}.$$

Show that a datum $Z(r = a, \theta_i, \phi_i)$ (vertical component of magnetic field at the Earth's surface) has A-matrix row:

$$A_{ij} = (l + 1) P_l^m(\theta_i) \cos m\phi_i$$

for a g_l^m model parameter and

$$A_{ij} = (l + 1) P_l^m(\theta_i) \sin m\phi_i$$

for an h_l^m model parameter. Here subscript $j(l, m)$ is a single label for the model parameters.

(b) Repeat for the other two cartesian components, X and Y.

8.6. Nonlinear inverse problems can be solved by linearising and iteration. In this case the rows of the A matrix are partial derivatives of the data with respect to the model parameters.

(a) In the geomagnetic problem above, total field data are nonlinear:

$$F = \sqrt{X^2 + Y^2 + Z^2}.$$

Obtain the partial derivatives with respect to g_l^m for some starting model $\boldsymbol{m}_0$ by using the chain rule:

$$\frac{\partial F}{\partial g_l^m} = \frac{X_0}{F_0}\frac{\partial X}{\partial g_l^m} + \frac{Y_0}{F_0}\frac{\partial Y}{\partial g_l^m} + \frac{Z_0}{F_0}\frac{\partial Z}{\partial g_l^m}$$

and substituting from question (8.5) for the X, Y, Z derivatives.

(b) Repeat for declination: $D = \tan^{-1}(Y/X)$.

8.7. You form a beam of P digitised seismograms $y_i(t)$; $i = 1, 2, \ldots, P$ by delay and sum:

$$B(t, \boldsymbol{\tau}) = \sum_{i=1}^{D} y_i(t - \tau_i),$$

where $\boldsymbol{\tau}$ is a P-length vector of time delays. The beam is made to fit (in a least-squares sense) to a signal $s(t)$. The number of data is the number of digitised time points in the seismograms. Find the elements of one row of the partial derivatives (A) matrix. See Section 11.1.

9

Continuous inverse theory

9.1 A linear continuous inverse problem

The object of this chapter is to explain how to invert a finite number of data for properties of a model that is a continuous function of an independent variable, and how the continuous theory relates to discrete inversion.

Consider inverse problems of the form

$$d_i = \int_a^b K_i(x) m(x) \mathrm{d}x, \tag{9.1}$$

where $\boldsymbol{d}$ is the usual D-length column matrix of data values d_i; $i = 1, 2, \ldots, D$ and $m(x)$ is a suitably well-behaved function of the independent variable x describing the model. The function K_i is called the *data kernel* for the ith datum. Equation (9.1) constitutes the *forward problem*; it provides the theory for computing the datum from the model. The inverse problem involves finding m from $\boldsymbol{d}$; it is linear because the relationship (9.1) is linear.

The illustrative example will be, as in Chapters 6 and 7, the determination of the Earth's density from measurements of its mass and moment of inertia. We now assume the density can be represented as a piecewise continuous function of radius. The forward problem is the same pair of equations (6.8) and (6.9):

$$x = \frac{r}{a} \tag{9.2}$$

$$M = 4\pi \int_0^a \rho(r) r^2 \mathrm{d}r \tag{9.3}$$

$$\frac{I}{a^2} = \frac{8\pi}{3a^2} \int_0^a \rho(r) r^4 \mathrm{d}r. \tag{9.4}$$

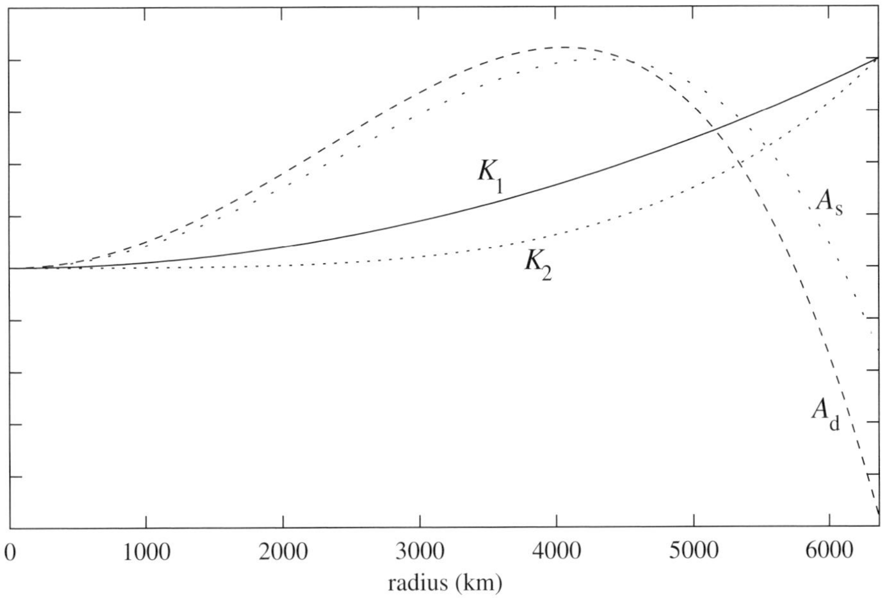

Fig. 9.1. Data kernels for mass and moment of inertia problem, from equations (9.5) and (9.6). A_d, A_s are the averaging functions designed to determine the density at half the Earth's radius using Dirichlet and quadratic spread functions respectively.

The data are scaled for algebraic convenience to the form

$$d_1 = \frac{M}{4\pi a^2}; \quad K_1(r) = x^2 \tag{9.5}$$

$$d_2 = \frac{3I}{8\pi a^4}; \quad K_2(r) = x^4. \tag{9.6}$$

The data kernel defines how the measurement samples the data; if $K(x) = 0$ for some value of x, equation (9.1) shows that the datum contains no information about the value of m at that point. In this particular example the data tell us nothing about the density at the Earth's very centre because both K_1 and K_2 vanish there. Plotting the data kernels for a particular inverse problem tells us which measurements are best for determining one particular aspect of a model. Figure 9.1 shows that both M and I sample the densities in the outer regions of the Earth more than those in the deep interior, but I samples the outer regions more strongly than M.

9.2 The Dirichlet condition

Each datum samples some average, weighted with the data kernel, of the model. We can determine other average properties of the model by taking linear combinations of the data. From equation (9.1)

$$\sum_{i=1}^{D} \alpha_i d_i = \int_a^b \sum_{i=1}^{D} \alpha_i K_i(x) m(x) \mathrm{d}x. \tag{9.7}$$

The $\{\alpha_i\}$ are called the *multipliers* and the combination of data kernels the *averaging function* $A(x)$,

$$A(x) = \sum_{i=1}^{D} \alpha_i K_i(x). \tag{9.8}$$

The linear combination of data now gives an average of the model weighted by the averaging function $A(x)$.

The multipliers must be chosen so that the averaging function yields some desired property of the model. Ideally we would like to know the model at every point x. That is clearly impossible as it would require an infinite number of data, but we proceed to see how well we can do with a finite number of data. The averaging function must be sharply peaked at the chosen value of x, which we call x_0. The Dirac delta function does exactly what we want because of its substitution property (Appendix 2):

$$\int_a^b \delta(x - x_0) m(x) \mathrm{d}x = m(x_0). \tag{9.9}$$

To this end, we choose multipliers according to the Dirichlet condition and minimise the mean squared difference between the averaging function A and the delta function

$$S = \int_a^b [A_\mathrm{d}(x, x_0) - \delta(x - x_0)]^2 \mathrm{d}x. \tag{9.10}$$

(The alert reader will notice this breaks one of the rules of the Dirac delta function (Appendix 2), that it should always multiply another function and never a distribution, but never mind, the problem disappears immediately.) The averaging function in (9.9) has been written explicitly as depending on x_0, the desired centre of the average; subscript 'd' indicates we are using the Dirichlet condition.

Differentiating with respect to each unknown multiplier in turn and setting the result to zero to obtain the minimum gives

$$\frac{\partial S}{\partial \alpha_j} = \int_a^b 2 \left[\sum_{i=1}^{D} \alpha_i K_i(x) - \delta(x - x_0) \right] K_j(x) \mathrm{d}x = 0. \tag{9.11}$$

Rearranging gives a set of D simple simultaneous equations for the unknown multipliers:

$$\sum_{i=1}^{D} \int_a^b K_i(x)K_j(x)\mathrm{d}x\alpha_i = K_j(x_0). \tag{9.12}$$

The matrix whose elements are the integrals on the left-hand side is sometimes called a *Gram* matrix

$$K_{ij} = \int_a^b K_i(x)K_j(x)\mathrm{d}x. \tag{9.13}$$

K is symmetric and positive definite, ensuring (9.12) always has a solution for the multipliers. After solving for the multipliers, the sum $\sum_{i=1}^{D} \alpha_i d_i$ gives the required estimate of $m(x_0)$ and (9.8) gives the averaging function.

Returning to the example of finding density from mass and moment of inertia, we seek an estimate of density at half the Earth's radius and put $x_0 = 0.5$ into equation (9.12). The Gram matrix elements are found by substituting from (9.5) and (9.6) for K_1 and K_2 into the integrals on the left-hand side of (9.12). These integrals are of simple powers of x. Equation (9.12) becomes

$$\begin{pmatrix} \frac{1}{5} & \frac{1}{7} \\ \frac{1}{7} & \frac{1}{9} \end{pmatrix} \begin{pmatrix} \alpha_1 \\ \alpha_2 \end{pmatrix} = \begin{pmatrix} \frac{1}{4} \\ \frac{1}{16} \end{pmatrix}, \tag{9.14}$$

which is readily solved to give

$$\alpha_1 = 10.391 \tag{9.15}$$

$$\alpha_2 = -12.797. \tag{9.16}$$

The density estimate is then

$$\bar{\rho}(a/2) = \frac{1}{a}\left(\alpha_1 \frac{M}{4\pi a^2} + \alpha_2 \frac{3I}{8\pi a^4}\right) = 7432 \text{ kg m}^{-3} \tag{9.17}$$

and the averaging function is

$$A_{\mathrm{d}}(x, x_0) = a(10.391x^2 - 12.797x^4). \tag{9.18}$$

A_{d} is plotted in Figure 9.1. It has a maximum somewhere near the right place $x = 0.5$ but is very broad, indicating poor resolution for determining the density at precisely the mid point. This is only to be expected with just two data. The estimate itself is lower than the best estimate of the density at this depth (10 330 kg m^{-3} from PREM) because the averaging function is biassed towards the outer half of the Earth. The solution for the density averaged with the known function $A_{\mathrm{d}}(x, x_0)$ is very precise and agrees with PREM almost exactly because PREM also fits these two data.

9.3 Spread, error and the trade-off curve

The resolution of a continuous inversion may be determined from the 'width' of the averaging function. It is often convenient to define resolution by a single number, such as the width of the averaging function at half the maximum height. The averaging function A_{d} in Figure 9.1 has a resolution of about half the Earth's radius, as might be expected from just two data. If, as is the case in many geophysical inverse problems, resolution is more important than error reduction, we should design the inversion to optimise resolution. This requires a suitable mathematical definition of resolution, which the Dirichlet condition fails to provide because it is impossible to evaluate the integral in (9.10).

Backus and Gilbert (1968) proposed a number of definitions of the *spread*, s, each with the dimensions of length and involving the product of the averaging function with a spread function with a minimum at the estimation point x_0. An example is

$$s^2 = \int_a^b (x - x_0)^2 A^2(x, x_0)\mathrm{d}x. \tag{9.19}$$

The function $(x - x_0)^2$ has a minimum at $x = x_0$, so when we minimise s^2 to find A the multipliers will adjust so as to make A large near $x = x_0$ and penalise values of A elsewhere. A normalisation condition is needed for A to prevent minimisation of s^2 leading to the trivial solution $A = 0$. The obvious choice is the area beneath A:

$$C = \int_a^b A(x, x_0)\mathrm{d}x - 1 = 0. \tag{9.20}$$

The averaging function corresponding to spread s and estimation point x_0 is found by minimising the integral on the right-hand side of (9.19) subject to the constraint (9.20). This is accomplished conveniently by the method of Lagrange multipliers, minimising the quantity

$$s^2 + \lambda C = \int_a^b (x - x_0)^2 A^2(x, x_0)\mathrm{d}x + \lambda \left[\int_a^b A(x, x_0)\mathrm{d}x - 1 \right],$$

where λ is the Lagrange multiplier. Substituting (9.8) into (9.19) and differentiating with respect to the multiplier α_k gives a set of linear equations:

$$\sum_{j=1}^{D} \int_a^b (x - x_0)^2 K_i(x) K_j(x)\mathrm{d}x \alpha_j = -\frac{\lambda}{2} \int_a^b K_i(x)\mathrm{d}x. \tag{9.21}$$

The constraint (9.20) becomes

$$\sum_{i=1}^{D} \alpha_i \int_a^b K_i(x)\mathrm{d}x = 1. \tag{9.22}$$

This equation is needed to solve for the Lagrange multiplier λ. The full set of equations is most easily solved by dividing (9.21) by $\lambda/2$, solving for the $2\alpha_i/\lambda$, then finding λ by using the normalisation condition (9.22).

The example of mass and moment of inertia will illustrate the method. Integrals of the data kernels enter all equations; these involve simple powers of x. The elements of the matrix involve integrals of higher powers of x. For example,

$$K_{11} = \int_0^1 (x - x_0)^2 x^4 \mathrm{d}x = \frac{1}{7} - \frac{x_0}{3} + \frac{x_0^2}{5}. \tag{9.23}$$

For $x_0 = 0.5$ the Gram matrix is

$$\mathsf{K} = \begin{pmatrix} 0.026\,19 & 0.2183 \\ 0.021\,83 & 0.018\,69 \end{pmatrix}. \tag{9.24}$$

The multipliers now become $\boldsymbol{\alpha} = (8.7266, -9.544)$ and the corresponding density estimate $\bar{\rho}(0.5) = 7339$ kg m^{-3}. The new averaging function, $A_s(x, 0.5)$, is plotted in Figure 9.1.

Comparing with the results from the Dirichlet condition (9.16), (9.18) shows quite close agreement in the multipliers and, in Figure 9.1, the averaging function. The density estimate is also similar, the difference arising because this estimate applies to a different average. Again, the average would agree with the PREM value. The spread itself is large: $s = 1.257$, or more than an Earth radius. We should not read too much into this, as s is a rather pessimistic estimate of 'width'.

For noisy data we construct a trade-off curve by minimising a combination of spread and error. The sum of squares of the errors is, from (9.1),

$$E = \sum_{i=1}^{D} \left[d_i - \int_a^b K_i(x) m(x) \mathrm{d}x \right]^2 \tag{9.25}$$

and the trade-off curve is constructed by choosing multipliers so as to minimise the combination

$$T(\theta) = S \cos\theta + E \sin\theta$$

with $0 < \theta < \pi/2$. As before, θ is the trade-off parameter, and this choice gives the full range of the trade-off curve from $\theta = 0$, where only the spread is minimised and we optimise resolution at the expense of fitting the data, to $\theta = \pi/2$, when only the error is minimised and no account is taken of resolution. The trade-off curve is a monotonic function of θ, showing that relaxing the fit will always improve the resolution, and relaxing the resolution will always improve the fit.

9.4 Designing the averaging function

The technique used in Section 9.2 to apply the Dirichlet condition can be generalised to obtain an approximation to any averaging function. We might, for example, wish to determine the average property throughout a layer, such as the mean density of the Earth's core, in which case we should use an averaging function equal to one in the core and zero outside it. Suppose the desired averaging function is $A_0(x)$. As before, we take a linear combination of the data to produce an averaging function given by (9.8). This time we choose the multipliers $\boldsymbol{\alpha}$ to minimise

$$E = \int_a^b [A(x) - A_0(x)]^2 \mathrm{d}x.$$

Differentiating with respect to each of the α_i in turn gives equations similar to (9.12), which reduces to the set of linear equations

$$\sum_{i=1}^{D} \int_a^b K_i(x) K_j(x) \mathrm{d}x \alpha_i = \int_a^b K_j(x) A_0(x) \mathrm{d}x. \tag{9.26}$$

The only difference with equation (9.12) is the right-hand side.

Consider again the example of finding density from mass and moment of inertia. Suppose we now wish to estimate the mass of the cores (outer + inner), M_c. The desired averaging function is proportional to x^2 for $x = x_c =< c/a$, where c is the core radius and a the Earth's radius, and zero outside. The right-hand sides of equation (9.26) become

$$\int_0^{x_c} x^4 \mathrm{d}x = \frac{x_c^5}{5} \tag{9.27}$$

$$\int_0^{x_c} x^6 \mathrm{d}x = \frac{x_c^7}{7}. \tag{9.28}$$

The core mass is then, from equations (9.5) and (9.6),

$$M_c = 4\pi a^3 \int_0^1 A_c(x) m(x) \mathrm{d}x = 4\pi a^3 \alpha_1 d_1 + \alpha_2 d_2 = \alpha_1 M + \alpha_2 \frac{I}{2a^2}. \tag{9.29}$$

Setting $x_c = 0.5470$ and solving the matrix equation gives $\alpha_1 = 0.4351, \alpha_2 = -0.5406$. The averaging function is plotted in Figure 9.2 alongside the ideal averaging function. $A_c(x)$ is too small and peaks at too large a radius. The estimate of

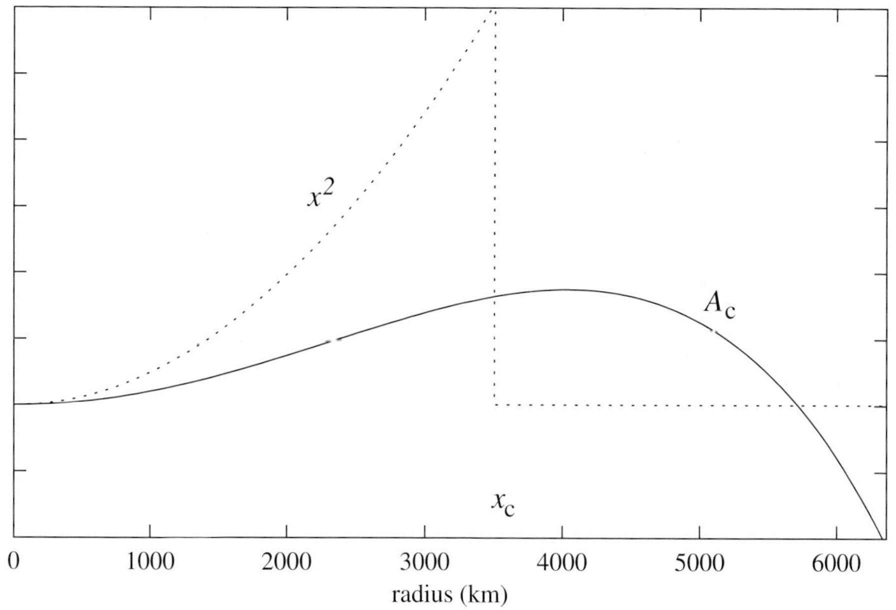

Fig. 9.2. Averaging function $A_c(x)$ designed to give an estimate of the mass of the Earth's core (solid line) and the desired averaging function $A_0(x)$.

the core mass is, from (9.29), 2.065×10^{24} kg, which compares well with the established value of 1.92×10^{24} kg. The result is therefore quite good for the mass, although the averaging function is very poor because we only have two data kernels to combine to form an averaging function with a discontinuity.

9.5 Minimum-norm solution

A minimum-norm solution may be obtained for the continuous inverse problem (9.1) as it was for the discrete under-determined problem in Chapter 7. The norm of the model is

$$|m| = \left[\int_a^b m^2(x) \mathrm{d}x \right]^{\frac{1}{2}}. \tag{9.30}$$

We first show that the minimum-norm solution is a linear combination of the data kernels, then determine the model from the coefficients multiplying the data kernels. Minimising the norm subject to satisfying the data is achieved by applying (9.1) as a set of constraints using the method of Lagrange multipliers. We minimise

$$E^2 = \int_a^b m^2(x) \mathrm{d}x + \sum_{i=1}^{D} 2\lambda_i \left[d_i - \int_a^b K_i(x) m(x) \mathrm{d}x \right], \tag{9.31}$$

where the $\{\lambda_i\}$ are Lagrange multipliers to be determined from the data constraints (9.1) (the factor 2 is included for later convenience). We make a small, arbitrary, change in the model to change it to $m(x) + \delta m(x)$ and set the change in E to zero:

$$\delta E = \int_a^b [2m(x) + \sum_{i=1}^{D} 2\lambda_i K_i(x)]\delta m(x)\mathrm{d}x = 0. \tag{9.32}$$

The result holds for any $\delta m(x)$ and therefore the integrand must be zero:

$$m(x) = \sum_{i=1}^{D} \lambda_i K_i(x). \tag{9.33}$$

This proves that the minimum norm solution is a linear combination of the data kernels $K_i(x)$.

It remains to determine the Lagrange multipliers by substituting (9.33) back into (9.1) to give

$$d_i = \sum_{j=1}^{D} \int_a^b K_i(x)K_j(x)\mathrm{d}x\lambda_j. \tag{9.34}$$

This is a matrix equation for the $\{\lambda_j\}$. We introduce vectors $\boldsymbol{d}$ and $\boldsymbol{\lambda}$. The Lagrange multipliers are then

$$\boldsymbol{\lambda} = \mathsf{K}^{-1}\boldsymbol{d} \tag{9.35}$$

and the minimum-norm solution

$$m(x) = \sum_{i=1}^{D}\sum_{j=1}^{D} (\mathsf{K}^{-1})_{ij} d_j K_i(x). \tag{9.36}$$

The minimum norm solution defines its own set of averaging functions. The estimate of m at a point x_0 is, from (9.36) and (9.35),

$$\begin{aligned} m(x_0) &= \sum_{i=1}^{D} (\mathsf{K}^{-1})_{ij} d_j K_i(x_0) \\ &= \int_a^b \sum_{i,j} (\mathsf{K}^{-1})_{ij} K_i(x_0) K_j(x) m(x)\mathrm{d}x. \end{aligned} \tag{9.37}$$

The averaging function is therefore

$$A(x, x_0) = \sum_{ij} (\mathsf{K}^{-1})_{ij} K_i(x_0) K_j(x). \tag{9.38}$$

The norm itself is obtained from (9.36):

$$|m|^2 = \int_a^b m^2(x)\mathrm{d}x = \int_a^b \sum_{i=1}^{D} \lambda_i K_i(x) \sum_{j=1}^{D} \lambda_j K_j(x). \tag{9.39}$$

In matrix notation, with (9.35), this becomes

$$|m|^2 = \boldsymbol{\lambda}^{\mathrm{T}}\mathsf{K}\boldsymbol{\lambda} = \boldsymbol{d}^{\mathrm{T}}\mathsf{K}^{-1}\boldsymbol{d}. \tag{9.40}$$

Application to the mass and moment of inertia problem is left as an exercise for the reader.

9.6 Discretising the continuous inverse problem

The great majority of geophysical inverse problems are continuous: they involve a model described by piecewise continuous functions. It is therefore surprising that the great majority of practitioners solve these inherently continuous problems by discretising them first, representing the model by a finite number of parameters before applying the methods of discrete inverse theory or even parameter estimation. There are two main reasons for this. First, the matrices involved in continuous inversion, like K in (9.35), are of order D, the number of data. This is usually very large, making numerical inversion of K impractical, although modern numerical techniques can provide good approximate solutions for very large linear systems. Discretising with a small number of parameters reduces the computational effort considerably. Secondly, the elements of K must be evaluated by integration. Only in rare instances can these integrals be done analytically, and often the matrix elements themselves must be computed numerically. Changing the inverse problem slightly, such as changing the norm, often involves a great deal of new work and programming effort, whereas the equivalent change for a discretised problem can usually be accomplished very simply.

To illustrate a straightforward discretisation, we approximate the model $m(x)$ by its value at a discrete set of equally spaced points $x_1, x_2, \ldots, x_P$, with $x_1 = a$ and $x_P = b$, and the integral in (9.1) by the trapezium rule to give

$$d_i = \frac{1}{2}[K_i(a)m(a) + K_i(b)m(b)] + \sum_{j=2}^{P-1} K_i(x_j)m(x_j)(x_j - x_{j-1}) + e_i. \tag{9.41}$$

This equation may now be expressed in the standard form (6.5) and we can proceed by the methods of discrete inversion outlined in Chapters 5 and 6. The purpose of the discretisation is simply to provide an accurate numerical representation of the continuous problem. It therefore has nothing to do with the inversion: it is an exercise in numerical analysis. P must be sufficiently large to provide an accurate representation of all the possible solutions to the inverse problem, not just some of the better-behaved ones. For this reason it is important to establish the general properties of the inverse problem, particularly those of existence and uniqueness, before discretising. Bad behaviour of the underlying mathematical problem is rarely obvious in the discretised problem because the inversion tends to find smooth, well-behaved solutions that fit the data. Two examples of this are given in Chapter 12 on geomagnetism. Both problems are inherently nonunique. In one (finding the geomagnetic field from directional measurements) the discretised problem retains a unique solution until $P \approx 60$, when common experience would suggest that perfectly adequate solutions would be obtained for $P \approx 10$; in the other (finding fluid motion in the core from secular variation) the nonuniqueness and damping allow you to find any result you want! These are not pathological cases, in my opinion they are quite the opposite: rare examples that are sufficiently simple for the real difficulties to be understood.

The task of discretisation is one of numerical analysis, and should be approached as such. I will not therefore attempt to deal with the subject here, save to indicate the considerations that determine the choice of discretisation. Point-wise discretisation, as described above, is the simplest. It can deal with most problems, with integrals performed by the trapezium rule and derivatives by finite differences. Another method is to expand in orthogonal functions. Legendre polynomials are used for polynomial expansions (see Appendix 5); spherical harmonics for spherical problems; and sometimes Bessel functions for cylindrical problems.

Finite element and spline approximations are computationally efficient: the critical decision here is *local* versus *global* representation. Polynomials, spherical harmonics, Bessel functions and the like are global approximations: the value of the approximation at one point depends on its value at every other point. Is this reasonable for the problem at hand? For the main geomagnetic field one could argue for a global representation because it originates in the core and we measure it at the Earth's surface where it is sensitive to contributions from all over the core surface. A local representation may be more appropriate for a seismic problem, where rays sample only a small part of the region around its path. Local representations also have the advantage of leading to sparse matrices, simplifying the computations.

Orthogonality is desirable in the representation. I give a common, generic representation here. We expand the model in a set of orthonormal functions (Appendix 5) $\phi_i(x);\ i = 1, 2, \ldots, P$ satisfying the conditions

$$\int_a^b \phi_i(x)\phi_j(x)\mathrm{d}x = \delta_{ij} \tag{9.42}$$

$$m(x) = \sum_{i=1}^{P} m_i \phi_i(x). \tag{9.43}$$

Equation (9.1) becomes

$$d_i = \int_a^b K_i(x) \sum_{i=1}^{P} m_j \phi_j(x) \mathrm{d}x + e_i. \tag{9.44}$$

This has the standard form of a discrete inverse problem (6.5) if we define a model vector whose components are the coefficients m_i and equation of condition matrix A to have elements

$$A_{ij} = \int_a^b K_i(x) \phi_j(x) \mathrm{d}x. \tag{9.45}$$

For orthonormal functions the right-hand side of (9.45) is simply the jth coefficient of the expansion of the data kernel $K_i(x)$ in the orthonormal basis.

Other examples of discretisation are to be found in the applications in Part III of the book, and in further reading. Note that discretisation simply reduces integral and differential expressions to algebraic forms; it can be used on nonlinear problems as well as linear ones, the only difference being that the resulting algebraic equations are themselves nonlinear.

9.7 Parameter estimation: the methods of Backus and Parker

In this section I give a brief overview of the general method of linear inference and parameter estimation developed by Backus (Backus, 1970b,c;d) with modifications by Parker (1977). Conventional parameter estimation relies on the assumption that errors stem entirely from the data. In most geophysical problems large errors arise from model inadequacy: the discretisation is not good enough to represent the real Earth. Franklin (1970) developed a formal approach to this problem by treating the model as a realisation of a stochastic process. This is not a very satisfactory approach, mainly because the Earth's properties are not randomly varying, but also because it involves some harsh assumptions about the statistical properties of the process. Rather than starting out by simplifying the model with a set of finite parameters, Backus sought to estimate the parameters from the continuous model.

We begin, as usual, with the data constraints (9.1), but now add the P parameters that are linearly related to the model:

$$p_k = \int_a^b m(x) P_k(x) \mathrm{d}x, \tag{9.46}$$

where $P_k(x)$ is a kernel defining the parameter. For example, if m is density, we could make p_1 the mean density of the core, in which case P_k would be x^2 inside the core and zero outside. Bounding the norm of the model,

$$\int_a^b m^2(x)\mathrm{d}x < M^2, \tag{9.47}$$

constrains all acceptable solutions to lie within an ellipsoid in function space (Appendix 5). The ellipsoidal form is important because it is a bounded function: it guarantees a finite volume of solutions. The data constraints (9.1) are linear and define a plane in the function space; the intersection of this plane with the ellipsoid defines the allowed solutions to the inverse problem. It is another ellipsoid. The parameters and their errors can then be estimated from the model. If the plane fails to intersect the ellipsoid the data are inconsistent with the model; there is no solution. This section contains a brief outline of the method and compares the results with the minimum-norm solution found in Chapter 7.

First we define some notation. Matrices P and H have elements that are integrals of the kernel functions $P_k(x)$ and $K_i(x)$:

$$P_{ij} = \int_a^b P_i(x)P_j(x)\mathrm{d}x \tag{9.48}$$

$$H_{ij} = \int_a^b P_i(x)K_j(x)\mathrm{d}x. \tag{9.49}$$

P and K (equation (9.13)) are square, positive definite, invertible matrices; H is rectangular $P \times D$. $\boldsymbol{P}$ is a P-length vector of functions $P_k(x)$; $i = 1, 2, \ldots, P$, and $\boldsymbol{K}$ is a D-length vector of functions $K_i(x)$. The set of functions $\tilde{P}_i(x)$; $i = 1, 2, \ldots, P$ that form the components of the vector $\tilde{\boldsymbol{P}}$ defined by

$$\tilde{\boldsymbol{P}} = \mathsf{P}^{-1}\boldsymbol{P} \tag{9.50}$$

are called the *dual basis* because they satisfy the orthogonality relation

$$\int_a^b \tilde{P}_i(x)P_j(x)\mathrm{d}x = \delta_{ij}, \tag{9.51}$$

which is easily verified by multiplying (9.50) by each function $P_i(x)$ and integrating. Note also that

$$\int_a^b \tilde{P}_i(x)\tilde{P}_j(x)\mathrm{d}x = (\mathsf{P}^{-1})_{ij}. \tag{9.52}$$

In this notation the minimum-norm solution (9.36) becomes

$$m(x) = \boldsymbol{K}^{\mathrm{T}}(x)\mathsf{K}^{-1}\boldsymbol{d}. \tag{9.53}$$

Substituting $m(x)$ from (9.53) into (9.46) gives the solution for the parameter vector $\boldsymbol{p}$

$$\boldsymbol{p} = \mathsf{H}\mathsf{K}^{-1}\boldsymbol{d}. \tag{9.54}$$

The expansion

$$m_p(x) = \sum_{i=1}^{P} p_i \tilde{P}_i(x) = \boldsymbol{p}^{\mathrm{T}}\tilde{\boldsymbol{P}}(x) \tag{9.55}$$

gives that part of the model represented by the parameters. It is easy to verify that (9.55) satisfies (9.46) using the orthogonality relation (9.51)

$$\int_a^b P_k(x)m_{\mathrm{p}}(x)\mathrm{d}x = \int_a^b \sum_{i=1}^{P} p_i \tilde{P}_i(x) P_k(x)\mathrm{d}x = p_k. \tag{9.56}$$

Result (9.54) may be compared with the discretised solution derived in Chapter 5, in which $m(x)$ is replaced by $m_{\mathrm{p}}(x)$ at the outset. In the present notation (9.1) becomes

$$\boldsymbol{d} = \mathsf{H}^{\mathrm{T}}\boldsymbol{p} \tag{9.57}$$

so H^{T} is the equations-of-condition matrix, and the unweighted least-squares solution is

$$\boldsymbol{p} = (\mathsf{H}\mathsf{H}^{\mathrm{T}})^{-1}\mathsf{H}\boldsymbol{d}. \tag{9.58}$$

This is different from (9.54).

Parker (1977) reports disappointing results for minimum-norm solutions because the *a priori* bound M that determines the size of the error ellipsoid can rarely be made sufficiently small. Furthermore, he finds it unreasonable to minimise the parametrised part of the model, which is presumably the part we think we know best. Instead he separates the parametrised part from the undetermined part m_*,

$$m(x) = m_{\mathrm{p}}(x) + m_*(x), \tag{9.59}$$

and minimises the norm $|m_*|$.

Proceeding as before we apply data constraints (9.1) by the method of Lagrange multipliers and minimise

$$E = \int_a^b [m(x) - m_{\rm p}(x)]^2 \mathrm{d}x + 2\boldsymbol{\lambda}^{\rm T}[\boldsymbol{d} - \int_a^b \boldsymbol{K}(x)m(x)\mathrm{d}x], \tag{9.60}$$

where $\boldsymbol{\lambda}$ is a vector whose components are Lagrange multipliers. We now change m by δm and $\boldsymbol{p}$ by $\delta \boldsymbol{p}$, then set the change in E to zero:

$$\int_a^b \{[m_*(x)\delta m(x) - \delta m_p(x)] - \boldsymbol{\lambda}^{\rm T}\boldsymbol{K}(x)\delta m(x)\}\mathrm{d}x = 0. \tag{9.61}$$

The term in $\delta m_{\rm p} = \delta \boldsymbol{p}^{\rm T}\tilde{\boldsymbol{P}}$ vanishes because the $\tilde{P}$ are orthogonal to m_*. To show this multiply (9.59) by $\boldsymbol{P}(x)$, integrate, and use (9.51) to give

$$\int_a^b m_*(x)\boldsymbol{P}(x)\mathrm{d}x = 0. \tag{9.62}$$

Premultiply by the matrix P^{-1} and use (9.50) to give the required result:

$$\int_a^b m_*(x)\tilde{\boldsymbol{P}}(x)\mathrm{d}x = 0. \tag{9.63}$$

The remaining terms in the integrand are all proportional to δm, and since the integral must vanish for all δm the integrand must vanish, leaving

$$m_*(x) = \boldsymbol{\lambda}^{\rm T}\boldsymbol{K}(x) \tag{9.64}$$

$$m(x) = \boldsymbol{p}^{\rm T}\mathsf{P}^{-1}\boldsymbol{P}(x) + \boldsymbol{\lambda}^{\rm T}\boldsymbol{K}(x). \tag{9.65}$$

The solution is a linear combination of both data kernels and parameter kernels. It remains to find the coefficients $\boldsymbol{\lambda}$, $\boldsymbol{p}$ using the data constraints.

Substituting (9.61) into (9.1) and using the definitions (9.48) and (9.49) gives

$$\boldsymbol{d} = \boldsymbol{p}^{\rm T}\mathsf{P}^{-1}\mathsf{H} + \boldsymbol{\lambda}^{\rm T}\mathsf{K}. \tag{9.66}$$

In manipulating these equations it is useful to recall that matrices K and P are square, symmetrical, and invertible. Furthermore $\mathsf{H}\mathsf{H}^{\rm T}$ is square and symmetric and will be invertible provided $P \leq D$. It follows that $\mathsf{HKH}^{\rm T}$ will be invertible if $P \leq D$. Rewrite (9.66) in terms of $\boldsymbol{\lambda}$:

$$\boldsymbol{\lambda} = \mathsf{K}^{-1}\boldsymbol{d} - \mathsf{K}^{-1}\mathsf{H}^{\rm T}\mathsf{P}^{-1}\boldsymbol{p}. \tag{9.67}$$

Substituting (9.64) into (9.62) gives

$$\int_a^b m_*(x)P_i(x)\mathrm{d}x = \sum_{j=1}^{D} \lambda_j \int_a^b K_j(x)P_i(x)\mathrm{d}x = 0, \tag{9.68}$$

or in matrix notation

$$\boldsymbol{\lambda}^{\mathrm{T}}\mathsf{H}^{\mathrm{T}} = \mathsf{H}\boldsymbol{\lambda} = 0. \tag{9.69}$$

$\boldsymbol{\lambda}$ may therefore be eliminated by premultiplying by H. Premultiplying (9.67) by H and rearranging gives

$$\mathsf{HK}^{-1}\boldsymbol{d} = \mathsf{HK}^{-1}\mathsf{H}^{\mathrm{T}}\mathsf{P}^{-1}\boldsymbol{p}. \tag{9.70}$$

Inverting the square matrices on the right-hand side in succession gives the solution

$$\boldsymbol{p} = \mathsf{P}(\mathsf{HK}^{-1}\mathsf{H}^{\mathrm{T}})^{-1}\mathsf{HK}^{-1}\boldsymbol{d}. \tag{9.71}$$

This solution is again different from the traditional parameter estimation (9.58). Parker (1977) finds the two are rather similar in his applications.

Now turn once again to the determination of the mean density of the Earth from mass and moment of inertia. In the dimensionless form given in Section 9.1 the mean density is

$$\bar{\rho} = \int_0^1 \rho(x)\mathrm{d}x, \tag{9.72}$$

so there is just one function $P_1(x) = 1$, the dual function $\tilde{P}_1(x)$ is also equal to 1, and so is the matrix P. K is the same as before and H is obtained by integrating the Ks to give

$$\mathsf{H} = \left(\frac{1}{3}, \frac{1}{5}\right). \tag{9.73}$$

Equations (9.54), (9.71) and (9.58) give the mean densities 4750, 6638, and 5262 kg m^{-3} respectively for the Backus, Parker, and least-squares estimates. The Backus estimate is minimum norm and so must be the smallest; Parker's is the largest because it minimises only the unrepresented part of the model; and the least-squares solution is biassed towards constant density, or mass divided by volume, which is 5515 kg m^{-3}.

The error bounds on these estimates are the main concern. These are discussed by Parker (1977) for problem (4), the determination of the mean densities of mantle and core from mass and moment of inertia. A recipe for the continuous inverse problem is given in Box 9.1.

Box 9.1. Summary: recipe for the continuous inverse problem

The following procedures are in addition to those listed for parameter estimation in Box 6.1, the under-determined problem in Box 7.1 and the nonlinear problem in Box 8.1.

(i) Establish as much as possible about the mathematical problem, particularly uniqueness and existence of solutions.
(ii) If no uniqueness proof is available, it may be possible to establish the null space (for an example see Section 12.8).
(iii) Next decide on whether to use the full apparatus of continuous inversion or to parameterise the problem and treat it as an under-determined, discrete inverse problem.
(iv) Most people choose to discretise and follow the procedures outlined in the previous chapters.
(v) Whatever you choose, a parametrisation is needed to effect a numerical solution. The choice of parametrisation is determined largely by the geometry of the problem and whether a local or global representation is appropriate.
(vi) Decide on choice of spread function or averaging function. This depends on what you wish to achieve by the inversion.
(vii) Form the Gram matrix and solve for the multipliers.
(viii) Use the multipliers to find the averaging function and model estimate.
(ix) Plot a trade-off curve to establish the required range of solutions.

EXERCISES

9.1. The mean density obtained by least squares in this chapter (5262) is different from that obtained in Chapter 6 (5383). Why?

9.2. Find an estimate of the derivative $g(x)$ of the function $f(x)$ (i.e. $g(x) = \mathrm{d}f/\mathrm{d}x$), where $f(x)$ is defined in the range [0,1], using the following method. Note that

$$f(x) = \int_0^x g(x')\mathrm{d}x' = \int_0^1 K(x, x')g(x')\mathrm{d}x',$$

where $K(x, x')$ is 1 when $x' < x$ and 0 otherwise. Use this form and the Dirichlet condition, described in Section 9.2, to estimate $g(0.25)$ by using the data below. Plot the averaging function. Compare your answer with the exact solution (the data are based on $f(x) = \sin \pi x$ and errors may be neglected). The Gram matrix in this case is singular and you should consider what to do about the null space.

$x =$	0.0	0.1	0.2	0.3	0.4	0.5	0.6	0.7	0.8	0.9	1.0
$f(x) =$	0.0	0.309	0.588	0.809	0.951	1.0	0.951	0.809	0.588	0.309	0.0

9.3. Repeat the estimation of the previous question by using the spread criterion described in Section 9.3. Calculate the spread and plot the averaging function.

9.4. Estimate the average of $g(x)$ between $x = 0.25$ and $x = 0.50$ in question 9.2 by designing the averaging function as described in Section 9.4. Compare your answer with the exact value. Plot the averaging function.

9.5. Obtain the minimum-norm solution for $g(0.25)$ by using the method described in Section 9.5. Plot the corresponding averaging function and compare your results to the solution for the Dirichlet condition obtained in question 9.2 and spread criterion in question 9.3.

9.6. Solve for $g(x)$ by first discretising $g(x)$ as a polynomial:

$$g(x) = a + 2bx + 3cx^2$$

and performing a standard discrete parameter estimation for the coefficients a, b, c. What answer would you expect if discretising with sines and cosines?

9.7. Obtain two further estimates of the average of $f(x)$ between $x = 0.25$ and 0.5 by using the parameter estimation methods of Backus and Parker described in Section 9.7. Compare with the solution to question 9.4.

Part III

Applications

10

Fourier analysis as an inverse problem

10.1 The discrete Fourier transform and filtering

The DFT of a time sequence $\{a_k;\ k = 0, 1, 2, \ldots, N-1\}$ was defined in (2.17) together with its inverse in (2.18):

$$A_n = \frac{1}{N}\sum_{k=0}^{N-1} a_k e^{-2\pi i n k/N} \tag{10.1}$$

$$a_k = \sum_{n=0}^{N-1} A_n e^{2\pi i k n/N}. \tag{10.2}$$

These two equations can be written in matrix form by defining vectors $\boldsymbol{a}$, $\boldsymbol{A}$, whose components are the time sequence and DFT samples respectively, and matrix D, whose elements are given by

$$D_{nk} = \frac{e^{-2\pi i n k/N}}{N}. \tag{10.3}$$

Equations (10.1) and (10.2) are then written in vector form as

$$\boldsymbol{A} = \mathsf{D}\boldsymbol{a} \tag{10.4}$$

$$\boldsymbol{a} = N\mathsf{D}^{\dagger}\boldsymbol{A}, \tag{10.5}$$

where† denotes the *Hermitian conjugate* or complex conjugate of the transpose (Appendix 4):

$$\mathsf{D}^{\dagger} = (\mathsf{D}^{T})^{\star}.$$

Equation (10.5) may also be written as $\boldsymbol{a} = \mathsf{D}^{-1}\boldsymbol{A}$ and the result is true for all equally spaced time samples $\boldsymbol{a}$ and corresponding frequency samples $\boldsymbol{A}$. It follows that D is invertible with inverse

$$\mathsf{D}^{-1} = N\mathsf{D}^{\dagger}. \tag{10.6}$$

Equation (10.5) constitutes an inverse problem for the DFT coefficients $\boldsymbol{A}$, the model, given the time sequence $\boldsymbol{a}$, the data. There are no errors and the inverse problem is equi-determined because the number of model parameters and number of data are both equal to N. So far this exercise in rephrasing the problem of finding the DFT as an inverse problem is not very interesting because all we need to do to solve (10.5) is invert the square matrix $N\mathsf{D}^\dagger$, and we do not even have to do that numerically because we have its analytical form. The inverse problem becomes more interesting when we consider the effects of errors in the time sequence and try to estimate a smaller number of frequency terms. The number of frequency samples is less than the number of time samples and the inverse problem posed by (10.5) is over-determined, and we can use the least-squares solution to reduce the error in each frequency estimate. This is a standard exercise in parameter estimation (with the slight complication that the model parameters are complex numbers rather than real ones) that can be solved by the methods given in Chapter 6.

The inverse problem becomes even more interesting when we try to estimate more than N spectral elements. Now the number of model parameters exceeds the number of data and the inverse problem is under-determined. The matrix $N\mathsf{D}^\dagger$ is rectangular and has no conventional inverse. The methods of inverse theory allow us to construct a number of generalised inverses to this matrix, for example the one giving the minimum-norm solution (7.5). There is a null space in the frequency domain that turns out to be related to the problem of aliasing. Consider the case of just one additional spectral estimate, A_N. Periodic repetition tells us that $A_N = -A_0$, and therefore the first row of the matrix $N\mathsf{D}^\dagger$ is equal to the last row. The one-dimensional null space is therefore given by all vectors parallel to $\boldsymbol{A}_0 = (1, 0, 0, \ldots, -1)$. The minimum-norm solution would therefore return spectral estimate $A_N = A_0$, the same as we would get from periodic repetition.

Continuous inverse theory provides a mechanism for estimating the Fourier integral transform (FIT) from discrete samples of the time function. We have to assume the time signal is band-limited, so the FIT is zero for $|\omega| > \Omega$. Equation (A2.2) in Appendix 2 provides the necessary inverse formula

$$f(t_k) = \int_{\Omega}^{\Omega} F(\omega)\mathrm{e}^{\mathrm{i}\omega t_k}\,\mathrm{d}\omega. \tag{10.7}$$

The samples $f(t_k)$; $k = 1, 2, \ldots, D$ constitute the data, $F(\omega)$ is the model, and $\exp \mathrm{i}\omega t_k$ are the data kernels. Solution proceeds by the methods of continuous inverse theory outlined in Chapter 9 (and see Parker (1977)). This approach is very flexible and opens up a number of interesting possibilities: for example, the time samples need not be equally spaced. The use of inverse methods in Fourier analysis has been explored by Oldenburg (1976).

The convolution, or filtering, process can also be cast in matrix form and considered as an inverse problem. Adding noise to equation (2.7) gives

$$c_p = \sum_{k=0}^{p} a_k b_{p-k} + e_p. \tag{10.8}$$

Writing out the convolution process in full shows it to be matrix multiplication:

$$\begin{aligned}
(a * b)_0 &= a_0 b_0 \\
(a * b)_1 &= a_1 b_0 + a_0 b_1 \\
(a * b)_2 &= a_2 b_0 + a_1 b_1 + a_0 b_2 \\
\cdots &= \cdots \\
(a * b)_{N-2} &= \cdots + a_{K-1} b_{N-2} + a_{K-2} b_{N-1} \\
(a * b)_{N-1} &= \cdots + a_{K-1} b_{N-1}.
\end{aligned} \tag{10.9}$$

The matrix form is obtained by defining vectors $\boldsymbol{c}$ and $\boldsymbol{a}$ containing the sequences c and a, and a rectangular matrix B whose elements are

$$B_{pk} = b_{p-k}. \tag{10.10}$$

B has the special form, called *Toeplitz*, with all elements on the same diagonal equal:

$$\mathsf{A} = \begin{pmatrix} b_0 & b_{-1} & b_{-2} & \cdots \\ b_1 & b_0 & b_{-1} & \cdots \\ b_2 & b_1 & b_0 & \cdots \\ \cdots & \cdots & \cdots & \cdots \end{pmatrix}. \tag{10.11}$$

Equation (10.8) then becomes the standard formula for a linear inverse problem

$$\boldsymbol{c} = \mathsf{B}\boldsymbol{a} + \boldsymbol{e}, \tag{10.12}$$

which may be solved by the methods described in Chapters 6 and 7. If a is a filter, inverse methods can be used to design the filter. The example of the Wiener filter is given in the next section, but inverse methods could be used to design other types, such as bandpass filters.

10.2 Wiener filters

In Section 4.3 we found three different methods to undo the convolution process defined by (10.8). A fourth, very common, method is to design a filter that does the best possible job in the least-squares sense. Such filters have been named after the mathematician Norbert Wiener. Suppose we have a sequence a of length K

and seek the filter f with a given length N which, when convolved with a, produces something close to a desired output sequence d. It will not usually be possible to obtain exactly the desired output because of the problems encountered in Section 4.3: sequence a may not be minimum delay, its amplitude spectrum may have zeroes, or the recursion formula may be unstable. The least-squares method guarantees minimum error for a filter of given length.

Let e be the $(N + K - 1)$-length sequence of differences between the convolution $f * a$ and the desired output d:

$$e_k = d_k - \sum_{n=0}^{N-1} f_n a_{k-n} \qquad k = 0, 1, \ldots, N + K - 1. \tag{10.13}$$

Equation (10.13) may be rearranged into the standard form (6.4):

$$\boldsymbol{d} = \mathsf{A}\boldsymbol{f} + \boldsymbol{e} \tag{10.14}$$

provided we identify the desired output sequence with the data, the filter coefficients with the model, and matrix A with the equations of condition matrix:

$$A_{kn} = a_{k-n}. \tag{10.15}$$

Standard methods of linear inverse theory therefore apply.

With no data weighting or damping the solution for $\boldsymbol{f}$ is

$$\boldsymbol{f} = (\mathsf{A}^{\mathrm{T}}\mathsf{A})^{-1}\mathsf{A}^{\mathrm{T}}\boldsymbol{d} \tag{10.16}$$

(cf. equation (6.26)). The elements of the normal equations matrix are

$$\begin{aligned}(\mathsf{A}^{\mathrm{T}}\mathsf{A})_{ij} &= \sum_k A_{ki}A_{kj} \\ &= \sum_k a_{k-i}a_{k-j} \\ &= (2N-1)\phi_{i-j},\end{aligned} \tag{10.17}$$

where ϕ is the autocorrelation of a defined in Section 4.2, equation (4.10). $\mathsf{A}^{\mathrm{T}}\mathsf{A}$ is a square, $K \times K$ Toeplitz matrix with, apart from a constant multiplier $(2N - 1)$, the autocorrelation values of a on each diagonal. It is symmetric because the autocorrelation is symmetric:

$$\mathsf{A}^{\mathrm{T}}\mathsf{A} = \begin{pmatrix} \phi_0 & \phi_1 & \phi_2 & \ldots \\ \phi_1 & \phi_0 & \phi_1 & \ldots \\ \phi_2 & \phi_1 & \phi_0 & \ldots \\ \ldots & \ldots & \ldots & \ldots \end{pmatrix}. \tag{10.18}$$

The remaining terms on the right-hand side of (10.16) are

$$(\mathsf{A}^{\mathrm{T}}\boldsymbol{d})_i = \sum_k a_{k-i} d_k$$
$$= (N + K - 1)c_i, \qquad (10.19)$$

where c is the cross correlation of a and d as defined in (4.9). Levinson's algorithm (Press *et al.*, 1992) can be used to invert Toeplitz matrices of order N in approximately N times fewer operations than the usual matrix inversion methods. The solution to (10.16) is therefore far quicker than the general linear inverse problem.

Applying constant damping to the inversion gives the solution

$$\boldsymbol{f} = (\mathsf{A}^{\mathrm{T}}\mathsf{A} + \lambda \mathsf{I})^{-1}\mathsf{A}\boldsymbol{d}. \qquad (10.20)$$

The normal equations matrix $\mathsf{A}^{\mathrm{T}}\mathsf{A}$ is made up of elements of the autocorrelation of a (10.18); the principal diagonal elements are all equal to ϕ_0 (apart from the constant factor $2N - 1$) and damping amounts to increasing the zero-lag autocorrelation coefficient. We may think of the modified autocorrelation as the original plus a spike of height λ at the origin. The Fourier transform of the autocorrelation is the power spectrum. The Fourier transform of a spike at the origin is a constant at all frequencies. In the frequency domain damping therefore amounts to adding a constant to every Fourier component. This is exactly what was done in the water-level method described in Section 4.3 in order to fill in holes in the spectrum of a and stabilise the deconvolution. Damping is therefore equivalent to the water-level method. Adding a constant to the diagonal of the normal equations matrix is sometimes called *prewhitening* because it has the effect of flattening the spectrum.

Clearly there is a close connection between the water-level method and the damping of inverse theory. The Fourier convolution theorem applies only to cyclic convolution, but provided the necessary zeroes are added to make the convolution theorem exact (see Section 2.3) it is possible to show complete equivalence of the two procedures as follows. Taking the Fourier transform of (10.20) and applying the convolution theorem gives

$$F(\omega) = \frac{D(\omega)A^\star(\omega)}{[A(\omega)A^\star(\omega) + \lambda]}. \qquad (10.21)$$

If $\lambda = 0$ the $A^\star$s cancel leaving division in the frequency domain: if $\lambda \neq 0$ we preserve phase information through the $A^\star$ in the numerator and divide by a prewhitened power spectrum with its holes filled, as was the case in the water-level method.

Zeroes are rarely added to turn the cyclic convolution into a discrete convolution, and the Wiener filter is usually chosen to be shorter than the record length: either can make a dramatic difference to the results for short time sequences. Wiener deconvolution must be regarded as superior in general to division in the frequency domain because it is based more soundly on the theory of least squares.

More general damping can be based on the estimated noise in the sequence a. This also has its counterpart in filter theory, where an estimated noise spectrum is used to flatten the spectrum. The estimated noise autocorrelation defines the model covariance $\mathrm{C}_\mathrm{m}^{-1}$ (Robinson and Treitel, 1980).

10.3 Multi-taper spectral analysis

No modern discussion of spectral analysis would be complete without mention of multi-taper methods, devised originally by Thomson (1982). We discussed spectral leakage in Section 3.2. Spectral leakage occurs when a harmonic signal is recorded for a finite length of time and energy spreads to neighbouring frequencies. A simple cut-off spreads the energy into side lobes that have central frequencies well away from the true frequency. The problem can be alleviated by tapering the data, multiplying the original time sequence by a function of time that decreases to zero at both ends. The original time sequence may be viewed as an infinite sequence multiplied by a boxcar taper with vertical cut-offs. Spectral leakage occurs into the side-lobes of the DFT of the boxcar (Figure 2.3). Tapers with smoother cut-offs, such as the cosine or Gaussian tapers, have smaller or no side lobes and therefore less spectral leakage. They also have broader central peaks, so the resolution of the central frequency is impaired relative to the boxcar. The trade-off between resolution and leakage is unavoidable; the only way to reduce spectral leakage while retaining resolution is to extend the length of the time sequence by making more measurements.

The traditional use of tapers is unsatisfactory in two respects. First, the choice of shape (cosine, Blackman–Harris, etc.) is somewhat *ad hoc*[1]. Secondly, tapering effectively throws away perfectly good data at the ends of the sequence by multiplying by less than unity. For a stationary time sequence this leads to loss of information. For non-stationary series, an earthquake seismogram for example where the signal has a distinct start time, we could extend the start and end of the time sequence to cover the head and tail of the taper, but this adds noise to the DFT without adding any signal and will therefore impair the spectral estimates.

[1] Section 3.2 does not do full justice to the properties and reasons for choice of particular tapers. For a fuller discussion see, for example, Harris (1978).

Both problems can be removed by designing *optimal tapers*. For the shape problem this requires a definition of 'optimal' in terms of the side lobes of the taper. While this gives a certain intellectual satisfaction in quantifying exactly what we mean by 'optimal' spectral leakage, many practitioners will see the exercise as simply replacing taper choice with an equally *ad hoc* optimisation formula and will prefer the flexibility of simply choosing a taper shape according to the problem at hand.

The loss of data problem is more fundamental and can be largely eliminated by the multiple taper method described below. First recall the DFT formulae (10.1) and (10.2). Subscript k measures time and n measures frequency. The physical quantities of time and frequency are determined by the sampling rate Δt, which gives the total length of record $T = N\Delta t$, the sampling frequency $\Delta \nu = 1/T$, and the maximum frequency $1/\Delta t$. Physical quantities have been eliminated from (10.1) and (10.2). Consider first the simple problem of measurement of a noise-free sequence composed of a single complex frequency, measured by subscript M, of complex amplitude μ

$$x_k = \mu \mathrm{e}^{2\pi \mathrm{i} Mk/N}. \tag{10.22}$$

We wish to estimate the amplitude μ without first knowing the frequency. We multiply by an arbitrary taper function w_k to form the new series y

$$y_k = w_k x_k \tag{10.23}$$

and take the DFT:

$$Y_n = \frac{1}{N} \sum_{k=0}^{N-1} w_k \mu \mathrm{e}^{2\pi \mathrm{i} Mk/N} \mathrm{e}^{-2\pi \mathrm{i} nk/N}. \tag{10.24}$$

We seek the optimal taper w_k that maximises the energy in the frequency band $M - P$ to $M + P$, where $2P$ defines the width of the band and therefore our design spectral resolution. We must normalise the taper coefficients, otherwise there would be no upper limit to the energy in the spectral band, and this is most easily done by maximising the energy inside the spectral band divided by the energy in the entire band, or

$$\lambda = \frac{\sum_{n=M-P}^{M+P} |Y_n|^2}{\sum_{n=0}^{N-1} |Y_n|^2}. \tag{10.25}$$

λ is always less than unity and was called the *bandwidth retention factor* by Park *et al.* (1987); $1/\lambda$ the *spectral leakage*. We choose M and P before determining the unknowns w_k implicit in (10.25).

Substituting from (10.1) for the Y_n, the numerator in (10.25) expands to

$$\sum_{n=M-P}^{M+P} \frac{|\mu|^2}{N^2} \sum_{k=0}^{N-1} w_k^\star e^{-2\pi i(M-n)k/N} \sum_{l=0}^{N-1} w_l e^{2\pi i(M-n)l/N}$$

$$= \frac{|\mu|^2}{N^2} \sum_{k=0}^{N-1} \sum_{l=0}^{N-1} w_k^\star w_l \sum_{n=-P}^{P} e^{2\pi i n(k-l)/N}. \qquad (10.26)$$

The sum over n is a geometric progression already done in Section 2.2 for the DFT of the boxcar taper; it sums to the ratio of sines. We now convert to matrix formulation by defining a vector $\boldsymbol{w}$ of taper values and matrix T with elements

$$T_{kl} = \frac{\sin\left[\pi (k-l)(2P+1)/N\right]}{\sin\left[\pi (k-l)/N\right]}. \qquad (10.27)$$

The numerator of (10.25) is then, in matrix form, simply the quadratic form $|\mu|^2/N^2 \boldsymbol{w}^\dagger \mathsf{T} \boldsymbol{w}$.

The denominator in (10.25) could be evaluated the same way, but it is much quicker to use Parseval's theorem (Section 2.3.8) and replace the sum in the frequency domain with the sum in the time domain

$$\sum_{n=0}^{N-1} |Y_n|^2 = \frac{\mu^2}{N^2} \sum_{k=0}^{N-1} |w_k|^2 = \frac{\mu^2}{N^2} \boldsymbol{w}^\dagger \boldsymbol{w}. \qquad (10.28)$$

The maximisation problem then reduces to finding the maximum of

$$\lambda = \frac{\boldsymbol{w}^\dagger \mathsf{T} \boldsymbol{w}}{\boldsymbol{w}^\dagger \boldsymbol{w}}. \qquad (10.29)$$

The matrix T is Hermitian, so the variational principle (Appendix 4) gives the stationary values of λ as the real eigenvalues

$$\mathsf{T} \boldsymbol{w}^{(p)} = \lambda^{(p)} \boldsymbol{w}^{(p)}. \qquad (10.30)$$

The optimal tapers are therefore the eigenvectors of the relatively simple matrix T. The vectors $\boldsymbol{w}^{(p)}$ are mutually orthogonal because they are the eigenvectors of a Hermitian matrix (Appendix 4). There is just one 'optimal' taper, that corresponding to maximum eigenvalue λ. All the eigenvalues are real and positive and may be ordered according to size: $\lambda_0 > \lambda_1 > \lambda_2 > \ldots$. All tapers corresponding to the larger eigenvalues will prove useful. The first four tapers are shown in Figure 10.1 and the corresponding eigenvalues are in the caption. Clearly the largest eigenvalue corresponds to the simplest taper function, but the remaining tapers correspond to only slightly smaller eigenvalues and therefore only slightly more spectral leakage.

The taper shapes are instructive. The zero-order taper, $\boldsymbol{w}^{(0)}$, is similar to the tapers considered earlier in Section 3.2. It is symmetrical about the centre, peaks

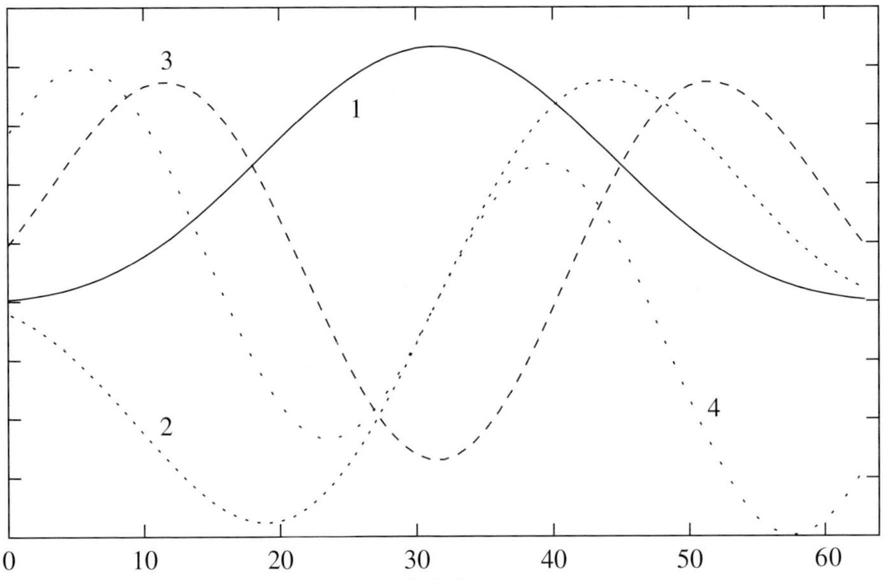

Fig. 10.1. The first four optimal tapers found by solving equation (10.29). Note that tapers 1 and 3 are symmetrical, 2 and 4 antisymmetrical. Taper 1 has least spectral leakage but does not use the extremes of the time series; higher tapers make use of the beginning and end of the record. The corresponding eigenvalues (bandwidth retention factors) are (1) 1.000 00, (2) 0.999 76, (3) 0.992 37, and (4) 0.895 84.

there, and tapers to zero at the ends. There is a symmetry about the higher-order tapers: those with even p are symmetrical about the centre, while those with odd p are antisymmetric. The taper with $p = 1$ has energy concentrated in the early part of the time sequence, and those with higher p are concentrated earlier still. The spectral estimate based on $\boldsymbol{w}^{(0)}$ will depend mainly on the central part of the time sequence, as is the case with conventional tapers, but estimates based on higher-order tapers will depend more on energy at the beginning and end of the sequence. This suggests combining the effects of several tapers to obtain an even better spectral estimate, the multi-taper approach, but first we must consider the effects of errors in the time samples.

In the presence of noise, equation (10.22) is modified to

$$x_k = \mu e^{2\pi i Mk/N} + e_k, \qquad (10.31)$$

where as usual $\boldsymbol{e}$ is the error vector. The usual trade-off between resolution and error demands that finer resolution in frequency be paid for by decreased accuracy of the estimate's amplitude. This suggests we should design the taper to optimise a combination of the spectral leakage and estimate variance. Multiplying (10.31) by the taper w_k gives

$$w_k x_k = y_k + w_k e_k. \tag{10.32}$$

Consider the case of uncorrelated, or white, noise (this does not lose generality since we can always transform the data to be uncorrelated and univariant, as described in Section 6.3). The noise energy in the tapered data is then proportional to $\sum_{k=0}^{N-1} |w_k|^2$.

We previously maximised the ratio of signal energy in the desired frequency band to energy in the entire signal in order to maximise bandwidth retention. This is obviously equivalent to minimising the inverse ratio, the total energy for fixed energy in the chosen band. We now wish also to minimise the ratio of the entire noise energy to the signal energy in the band, the same quantity. For uncorrelated errors the above procedure therefore minimises the noise in the band. The tapers are unchanged. Correlated noise leads to different tapers; a slightly different quantity must be minimised.

Now suppose we wish to estimate the amplitude of the signal, μ. This is obtained from the DFT of the tapered data y_k. Rearranging equation (10.32) and taking the transform gives

$$Y_n = \mu W_{n-M} + N_{\mathrm{n}}, \tag{10.33}$$

where N_{n} is the DFT of the tapered noise $w_k e_k$ and we have used the shift theorem (Section 2.3.2) to obtain the DFT of $\exp 2\pi \mathrm{i} Mk/N w_k$ as W_{n-M}. Then μ may be obtained as a least squares fit to these equations.

It is now clear that using just one taper to find μ fails to use all the information in the data. The other tapers, although not having quite as good spectral leakage properties as the optimal taper, use data from the ends of the record that were eliminated by the first taper. We can write down similar equations to (10.33) for several high-order tapers with good leakage properties:

$$Y_n^{(p)} = \mu W_{n-M}^{(p)} + N_{\mathrm{n}}^{(p)}; \quad p = 0, 1, 2, \ldots. \tag{10.34}$$

This is a larger set of equations to solve for the same μ, the amplitude. We expect the solution to be more accurate than the estimate for just the optimal taper because it uses more of the time series.

Finally note that the noise in (10.34) is correlated because the original uncorrelated noise has been multiplied by a number of different tapers. We therefore need to use the least-squares solution (6.59) that takes the covariance matrix of the errors into account. The covariance matrix of the tapered noise spectra is proportional to

$$(\mathsf{C}_{\mathrm{e}})_{ij} = \sum_{n=0}^{N-1} W_n^{(p)} W_n^{(p)}. \tag{10.35}$$

Multi-taper methods have been used for more than simply making spectral estimates. An excellent account is in Percival and Walden (1998). In geophysics they have been used to estimate amplitudes of decaying free oscillations (Park *et al.*, 1987); attenuation (Masters *et al.*, 1996); wavelet estimation on a seismic section (Walden, 1991); spectral analysis of high-frequency (Park *et al.*, 1987); and multi-dimensional (Hanssen, 1997) spectral estimation; coherences of magnetotelluric time series (Constable *et al.*, 1997); spectra of climatic time series (Lall and Mann, 1995) and of stratigraphic successions (Pardoiguzquiza *et al.*, 1994); the retrograde annual wobble (King and Agnew, 1991); and the acoustic signal of thunder from seismic records (Kappus and Vernon, 1991). Software is available from Gabi Laske's web site (Appendix 7).

11

Seismic travel times and tomography

11.1 Beamforming

Arrays of seismometers are used for two principal purposes: to increase the signal–noise ratio by stacking many records together, and to measure the direction of the incoming wavefront. Relative arrival times need to be known accurately. Until the recent advent of long-wave radio, then satellite timing, this could be achieved only by telemetering every station to a central recorder and clock. The term *network* was used for a group of stations with independent timing in order to distinguish it from an *array* with central timing. Timing on a network was usually too poor to allow most of the array techniques to work. Many seismic arrays were set up to detect nuclear explosions, for example *NORSAR* in Norway and *Yellowknife* in Canada, and these have been used to do much good seismology unrelated to monitoring nuclear tests. Marine seismic surveys use extensive arrays of geophones towed behind ships to record airgun sources. Since the late 1980s it has been possible to deploy portable seismometers, each with their own clock, in temporary arrays to study particular areas, a great step forward in the use of seismology in earthquake and tectonic studies. Nowadays we talk of a global array of seismometers, and do array processing on the Earth's spherical surface.

Stacking requires alignment of the onset of the same seismic phase, or wave, at every instrument on the array. Once the right arrival has been identified and the right time delay introduced, summing all the traces increases the signal-to-(random) noise ratio by a factor of the order of the square root of the number of traces. It is often used to identify signals that are otherwise buried in the noise. The simplest way to identify the onset time is to read the records by hand, but this is a very time-consuming practice and impractical for large arrays. If the incoming wavefront is planar, a simple formula gives the relative arrival times as functions of the position of each station. The normal to the wavefront, or assumed ray path, is defined by two angles, the azimuth and dip. The dip is usually replaced by the *slowness*, the inverse of the speed with which the wavefront travels across horizontal ground. Azimuth and slowness may be estimated by least-squares inversion using the relative arrival times as data. They yield the direction and distance from the array to the source, assuming the source to be on the Earth's surface. The largest sources of error are caused by indistinct picks and delays introduced by local anomalies beneath the arrays. Residuals to the plane wavefront arrival times can be inverted to map the structure beneath the array using methods such as that of Aki *et al.* (1977), discussed in Section 11.2 below.

Very large arrays demand an automatic way to determine the time delays. There are two approaches: cross correlate the traces and pick the maximum, as described in Section 4.2 (see also vanDecar and Crosson (1990)), or invert for the time delays that maximise the energy in the stack. The second approach is described here as an example of an inverse problem that contains several of the features discussed in Part II; it is described in full in Mao and Gubbins (1995).

Suppose we have N seismic traces $\{y_i(t);\ i = 1, 2, \ldots, N\}$, where t is the time. We produce a *beam*, y_{B}, by delay-and-sum,

$$y_{\mathrm{B}}(t) = \sum_{i=1}^{N} w_i y_i(t + \tau_i), \tag{11.1}$$

where w_i is a weight and τ_i the time delay for the ith trace. We use a weighted sum rather than a simple addition because some traces may be noisier than others and we wish to reduce the effect on the beam. We now fit the beam to a chosen reference trace $y_0(t)$ by least squares. The original seismic traces are discretised in time. The equations of condition are therefore

$$y_0(t_k) = \sum_{i=1}^{N} w_i y_i(t_k + \tau_i) + e_k, \tag{11.2}$$

where k runs over the entire recording time. We choose a time window running from time t_1 through time t_M and minimise the sum of squares of differences between the reference trace and the beam within this window,

$$E^2 = \sum_{k=1}^{M} [y_0(t_k) - y_{\mathrm{B}}(t_k)]^2, \tag{11.3}$$

by adjusting the unknown weights and delays that appear in y_{B}. The optimum delays are estimates of the relative arrival times.

Equation (11.2) defines the inverse problem we wish to solve. The 'data' in this problem are, rather strangely, discretised values of the reference trace, which we define properly later. For the time being we assume we know this trace and that it has uniform errors. This defines the data array

$$d_k = y_0(t_k). \tag{11.4}$$

We normalise each trace to unit maximum; this does nothing to the solution as we are going to assign each trace an unknown weight. The unknown delays and weights are arranged into arrays $\boldsymbol{\tau}$ and $\boldsymbol{w}$; the model vector may then be written in the partitioned form $\boldsymbol{m} = (\boldsymbol{\tau}|\boldsymbol{w})^{\mathrm{T}}$.

The inverse problem (11.2) is linear in $\boldsymbol{w}$ but nonlinear in $\boldsymbol{\tau}$; we solve by linearisation as described in Section 8.3. The partial derivatives matrix for the weights comprises simply the traces themselves:

$$A^w_{kj} = \frac{\partial y_{\mathrm{B}}(t_k)}{\partial w_j} = y_j(t_k + \tau_j). \tag{11.5}$$

The partial derivatives matrix for the time delays is obtained by differentiating (11.2) and involves the time derivatives of the traces, denoted $\dot{y}(t)$:

$$A^\tau_{kj} = w_j \dot{y}(t_k + \tau_j). \tag{11.6}$$

The full equations of condition matrix may be written in partitioned form as

$$\mathsf{A} = (\mathsf{A}^\tau | \mathsf{A}^w). \tag{11.7}$$

We need a starting model, which for this problem must be chosen with care because of the danger of cycle skipping (p. 134), where a trace may skip an entire cycle and give a time delay that is in error by one full period. The starting model for the delays is computed from a first approximation using a plane wavefront followed by a grid search of time samples, shifting each trace back and forth by several time samples, until the beam energy reaches a maximum. The starting model for the weights is simply a set of ones.

Model damping is essential because there are two distinct classes of model parameter with different dimensions. Denote the damping constants for the weights by the components of the vector $\boldsymbol{\lambda}^w$ and for the delays by $\boldsymbol{\lambda}^\tau$. When $\boldsymbol{\lambda}^w$ is large and dominates $\boldsymbol{\lambda}^\tau$ the weights are constrained to stay at unity (the 'penalty method', p. 119); when $\boldsymbol{\lambda}^w$ dominates the delays are constrained to stay at the starting values. The overal level of the ratio of $\boldsymbol{\lambda}^w$ to $\boldsymbol{\lambda}^\tau$ must be found by trial and error.

Finally we return to the choice of reference trace. We could choose one of the traces, perhaps the one that looks least noisy, but this is obviously not optimal. Instead, Mao and Gubbins (1995) chose each of the traces in turn to form a set of misfits,

$$E_j^2 = \sum_k [y_j(t_k) - \sum_{i \neq j} w_i y_i(t_k + \tau_i - \tau_j)]^2, \tag{11.8}$$

and minimised a sum of all the misfits:

$$E^2 = \sum_j^N E_j^2. \tag{11.9}$$

This scheme avoids the need to choose a specific reference trace and works well in practice, being insensitive to a few bad traces. The weighting was found to work well in eliminating noisy traces, giving solutions with $w_i \approx 0$ for traces from faulty instruments.

Mao and Gubbins (1995) give two applications to real data. The first was from a surface, vertical-component, linear-array study performed in 1992 by British Coal. Geophones were put on land with a 10 m spacing. Shots were fired at different depths in a borehole. Here we use 68 upgoing *P* wave records from the first shot at 50 m depth (Figure 11.1). The 64th record is at the head of the borehole. It is obvious that some arrival times deviate from a hyperbola; this is caused by variations in elevation, weathering thickness, and seismic velocity. The initial solution was obtained by a linear relationship between arrival times and offsets. The initial trace-to-trace alignment shows obvious residual statics (Figure 11.1). After six iterations the inversion converged to a reasonable solution. The trace-to-trace alignment is shown in Figure 11.1; the residual statics have clearly been minimised.

Their second dataset comprised seismograms from a broadband, three-component array operated by the University of Leeds in the Tararua Mountain area of North Island, New Zealand (Stuart *et al.*, 1994). They selected a surface wave magnitude 6.7 event with epicentral distance 11.8° that occurred in the Kermadec Islands at 16:23 on 1 November 1991 (the event is also shown in Figure 1.1). The waveforms are complex (Figure 11.2) and strongly influenced by fine-scale structure associated with subduction of the Pacific Plate beneath the array, providing a very severe test of the method. They first bandpass-filtered the records from 1 to 2 Hz (label 'd' in Figure 11.3). Seven three-component seismograms were analysed. The time delay was referred to the top trace (ltn1).

The initial solution vector was obtained by assuming a plane wavefront with fixed azimuth and apparent slowness. The iterative procedure converged provided the starting solution was not off by more than 10° in azimuth and 0.02 s km^{-1} in slowness. This convergence criterion will, of course, depend in general on the array size and source location. Figure 11.3 shows inverted results for vertical, East–West and North–South components with initial solution calculated from an azimuth of 40° and slowness 0.1 s km^{-1}. The method converged after five iterations: the trace-to-trace comparison shows alignment of the *P* wave signal in all three components.

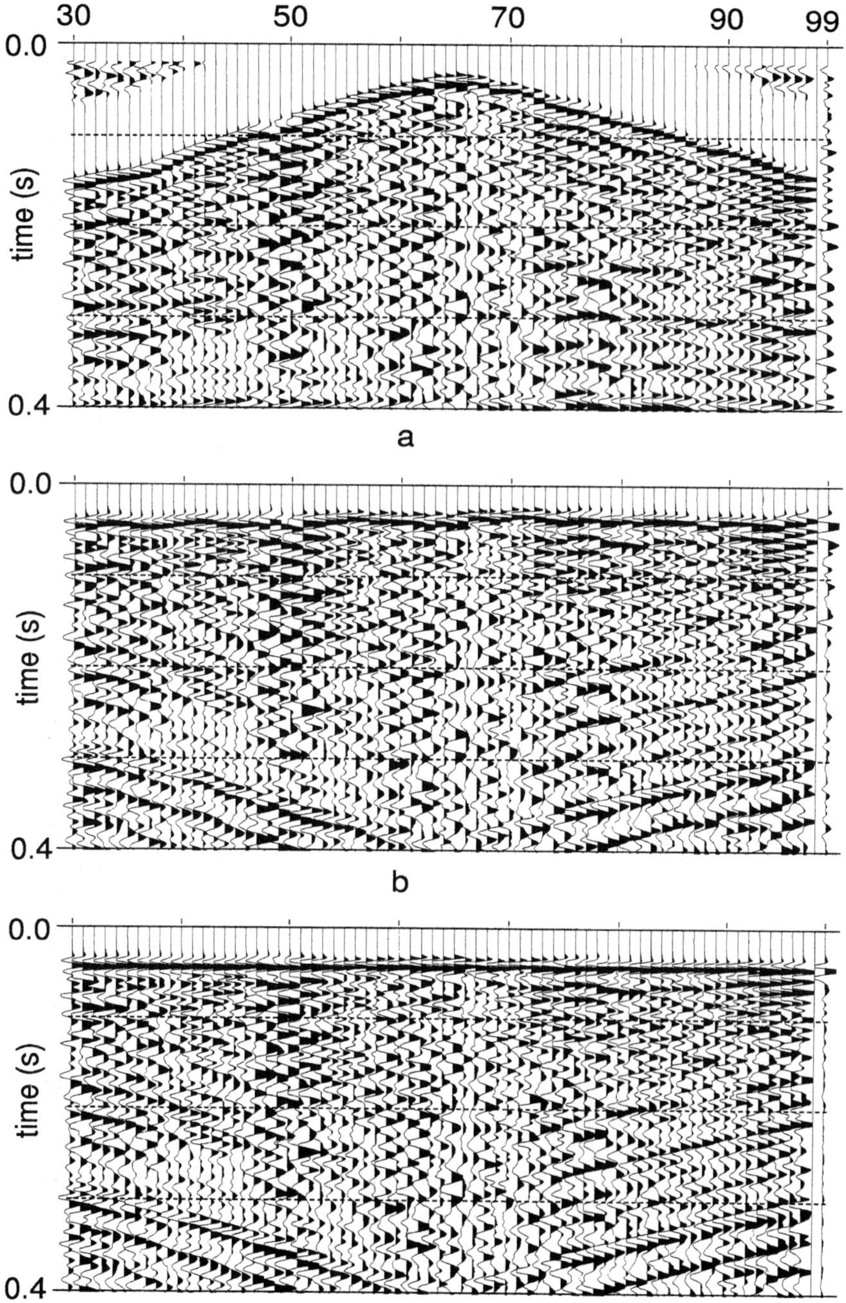

Fig. 11.1. (a) Array records from a borehole shot to 68 geophones on land with 10 m spacing. (b) Results from the initial model, found by a grid search. Trace 99 is the beam formed by a simple stack with unit weighting. (c) Results of the inversion. Reproduced, with permission, from Mao and Gubbins (1995).

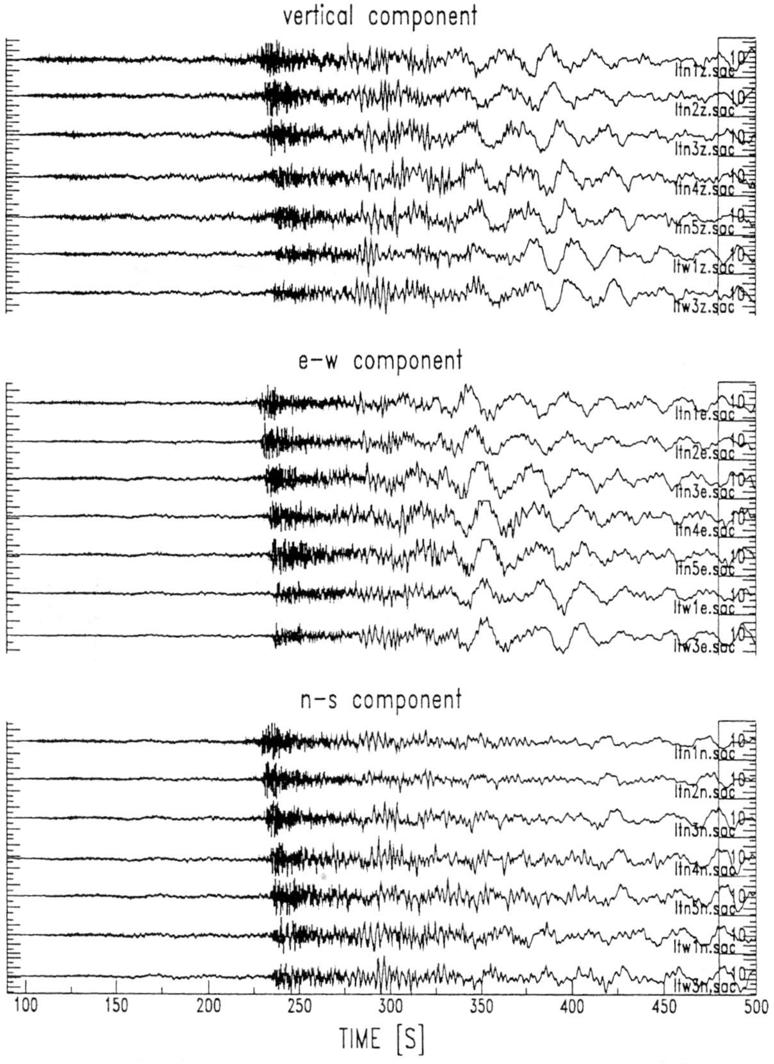

Fig. 11.2. A regional event (1 November 1991 $16^h23^m22^s$, 30.255° S 177.981° W, Ms 6.7) at the broadband, three-component array operated by the University of Leeds, UK, in the Tararua Mountain area of North Island, New Zealand. The time axes do not correspond to the origin time of the earthquake. Reproduced, with permission, from Mao and Gubbins (1995).

A similar analysis of the *S* wavelet was implemented. A time window of 4 s was taken from 228 s to 232 s. The initial solution was calculated with azimuth 40° and apparent slowness 0.21 $\mathrm{s\,km^{-1}}$. The *S* wavelet is clearly aligned on the East–West component, which is close to transverse for this wave. It does not seem to align as well on the North–South component, and worse still on the vertical,

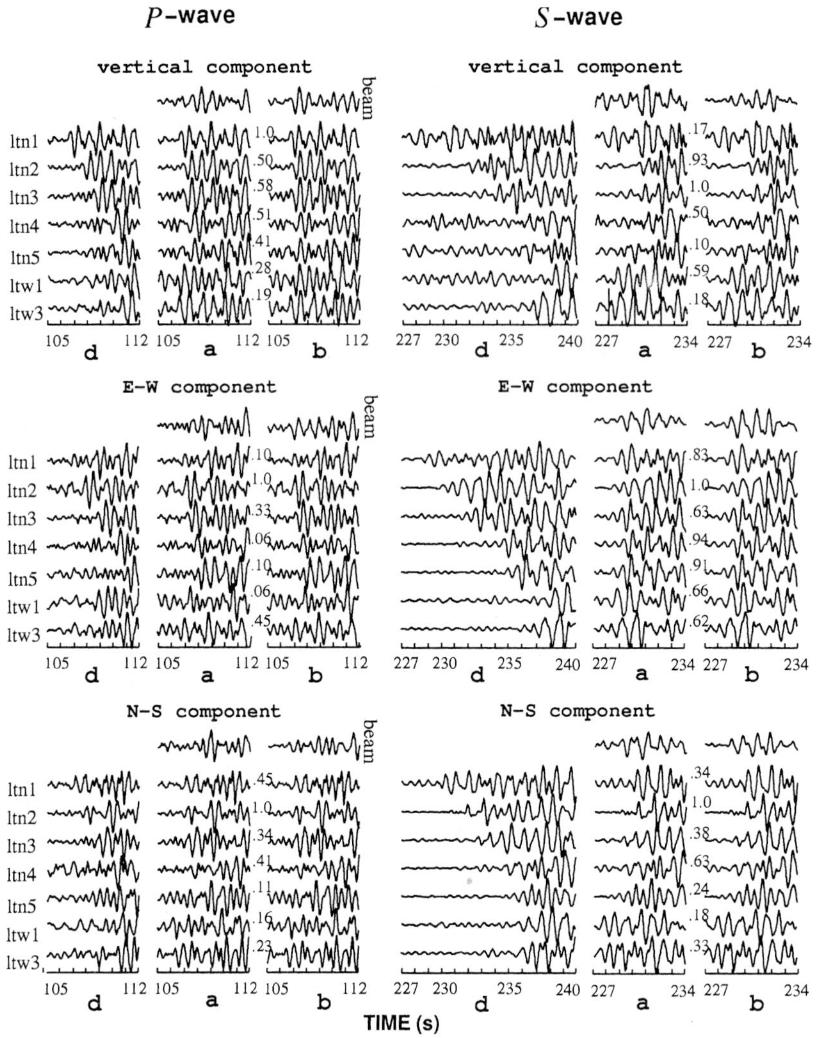

Fig. 11.3. Results of the filtered *P* and *S* wavelets in the frequency band 1 to 2 Hz for the regional event shown in Figure 11.2. Label 'd' stands for data, 'a' for initial model and 'b' for inverted model. The numbers at the left of each trace of inverted model are stacking weights. An *SP* converted phase is present in the vertical and NS components of ltw1 and ltw3, close to the *S* wave. It is suppressed in the *S* wave beams. Reproduced, with permission, from Mao and Gubbins (1995).

particularly for stations ltw1 and ltw3, which lie to the west of the main arm of the array. Mao and Gubbins (1995) believe this to be caused by conversion of the incoming *S* wave to a *P* wave at an interface beneath the array, probably the upper surface of the subducted Pacific Plate. Since *S* waves and *SP* converted waves have different paths and arrival times at each station, the *SP* phases are suppressed in the beams when the S wavelet aligns. For the P wave the calculated back azimuth was 35° and apparent slowness 0.103 s km^{-1}; for the S wave they are 47° and 0.209 s km^{-1}. Both azimuths are significantly larger than the source–receiver azimuth (29°) because of lateral heterogeneity in this complex subduction zone.

11.2 Tomography

Arrival times are the basic measurement of seismology. Interpretation of the full waveform of a seismogram is a relatively recent innovation, following the development of better seismometers, digital recording, and the computing power needed to compute the complex waveforms produced by even simple Earth structures. Travel times are predicted with remarkable accuracy by geometrical ray theory, simply by applying Snell's law progressively along the ray path. Travel times are in fact an average of the inverse of the velocity, sometimes called the *slowness* (not to be confused with the slowness of a wave across an array). Travel times can therefore be inverted, in principle at least, for the Earth's seismic velocity. In geophysics the approach used to be called travel-time inversion; the word tomography was coined later for medical applications and is now used almost universally throughout geophysics. Tomography comes from the Greek word for slice; the method images slices of the target material by moving the source of sound waves and detectors around to sample different slices. Seismic tomography suffers relative to medical tomography in two ways: the rays are not straight because of strong refraction within the Earth, and we are restricted to placing our sources and receivers on the Earth's surface or using earthquake sources at poorly known locations. In medical applications it is possible to surround the whole body with sources and receivers. Refraction turns out to be an annoyance rather than a fundamental difficulty; the surface restriction for sources and receivers poses much more serious problems, and the inverse problem can be highly under-determined.

Examples of seismic tomography are legion and cover a wide range of geometries and applications. The earliest form of tomography was to invert the travel times tabulated as a function of distance for the seismic velocity (P or S according to the wave measured) as a function of depth. If the travel time is available as a continuous function of distance it is possible to solve for the velocity provided it increases continuously with depth. This result was proved by Herglotz and Wiechert by solving an Abel integral equation (Bullen and Bolt, 1985). The result applies in both plane and spherical geometry. The approach is not the best way to invert real data that fail to produce a well-defined continuous function of travel time against distance, but the analytic solution is vital in highlighting problems arising from low-velocity zones: these create invisible layers which cannot be explored by surface measurements. The inverse problem has, in effect, a null space.

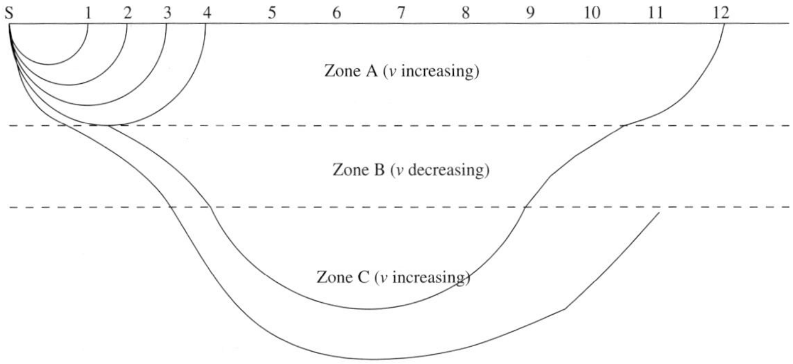

Fig. 11.4. One-dimensional seismic tomography.

The process of travel time inversion for a layered medium is shown in Figure 11.4. The shortest ray, S1, gives an estimate of the seismic velocity in the shallowest depths. This can be used to strip out the time that the next deepest ray, S2, spends in the shallowest region, leaving an estimate for the seismic velocity in the second shallowest region, and so on. Ray S4 just reaches the zone where seismic velocity begins to decrease with depth. Rays in zone B are refracted downwards, creating a shadow zone on the surface. No rays reach stations 5–11. Ray S12 has reached the depth of zone C, where the seismic velocity increases again, and the ray returns to the surface. Clearly we have no information about the seismic velocity in zone B, the low-velocity layer. We can proceed to determine velocities throughout zone C as we did for zone A. If a continuous time–distance curve is first constructed by interpolation we can determine the seismic velocity as a continuous function of depth throughout the zones where seismic velocity increases with depth.

Aki *et al.* (1977) took an important step in using travel times to determine laterally varying structures beneath seismic arrays, using the excellent dataset from the *NORSAR* seismic array in southern Norway. Dziewonski (1984) performed the first inversion to give lateral variations on the global scale. In exploration geophysics, a great deal of effort has gone into cross-borehole tomography, in which shots in one borehole are recorded in another and delays in arrival times are interpreted in terms of structure between the boreholes; and into reflection–transmission tomography, which combines the inversion of reflection times for the depths to interfaces with delays introduced by lateral variations in the intervening layers. Despite these efforts, exploration seismology is still dominated by reflection work because of its higher resolution.

The travel time between two points A and B is given by the path integral,

$$T_A^B = \int_A^B \frac{dl}{v(x, y, z)} = \int_A^B s(x, y, z)\,dl, \tag{11.10}$$

where the integral follows the ray path from A to B and $s = 1/v$ is the slowness. This equation poses a continuous inverse problem in its basic form: the data are a set of travel times between a set of pairs of points A and B and they are to be inverted for the model, the velocity of the intervening structure, v. Despite appearances, the inverse problem in terms of s is nonlinear because the integral follows the ray path and the ray path depends on the slowness.

Linearisation of (11.10) uses *Fermat's principle* of least time, which asserts that the travel time will be stationary with respect to changes in the ray path provided the integral is taken over a true ray path. In the linear approximation a small change in the slowness δs, will produce a small change in travel time δT_A^B, according to the integral

$$\delta T_A^B = \int_A^B \delta s(x, y, z) \mathrm{d}l \tag{11.11}$$

where, because of Fermat's principle, the integral is taken over the same path as originally. This is an important simplification; without Fermat's principle we would have to add an additional term to (11.11) to allow for the change in travel time caused by the change in ray path. Note that changing an end point A or B by a small amount does change the travel time to first order. Times of reflection are therefore changed to first order by shortening or lengthening of the ray path as the depth of the reflector changes with the change in structure. This is why reflection seismics are more sensitive to structure than transmission tomography.

Most studies proceed by parametrising the slowness with suitable functions $c_j(x, y, z)$,

$$\delta s(x, y, x) = \sum_{j=1}^{P} m_j c_j(x, y, z), \tag{11.12}$$

to convert the continuous form to the standard discrete linear form

$$T_i = \sum_{j=1}^{P} \int_{\Gamma_i} c_j(x, y, z) \mathrm{d}l \; m_j, \tag{11.13}$$

where now Γ_i denotes the ray path of the ith travel time. The partial derivatives matrix has elements given by the ray path integrals of the basis functions in the parametrisation

$$A_{ij} = \frac{\partial T_i}{\partial m_j} = \int_{\Gamma_i} c_j(x, y, z) \mathrm{d}l. \tag{11.14}$$

Typical basis functions used include uniform blocks (Aki *et al.*, 1977), spherical harmonics for global studies (Dziewonski, 1984), and splines (Mao and Stuart, 1997).

The nonlinear problem is solved by iteration from a starting model. The partial derivatives must be calculated at each step using (11.14) by first computing the ray path through the existing structure, then performing the path integral for each of the basis functions $c_j(x, y, z)$. In terms of computation the time-consuming step is calculating the ray path, and the form of parametrisation is best chosen to facilitate this part of the inversion. Most inversions have used only one step, effectively assuming a linear approximation. There are also many examples of nonlinear inversions that include ray tracing at each step (although I remain unconvinced of the need for extra ray tracing in many cases).

Solution of the one-dimensional problem, in which the slowness is a function only of the depth, is illustrated graphically in Figure 11.4. Low-velocity zones prevent solution. This is one of the few tomographic problems to have been tackled using continuous inverse methods (Johnson and Gilbert (1972)). Each ray path is identified by a parameter $p = \sin I/v$, where I is the angle of incidence, and p, by Snell's law of refraction, is constant along the ray path. A little work shows the ray path $z(x)$ satisfies the differential equation

$$\frac{dx}{dz} = \frac{ps}{\sqrt{1 - p^2 s^2}} \tag{11.15}$$

and the travel time integral is

$$T(x) = 2\int_0^{z_0} \frac{ps}{\sqrt{s^2 - p^2}} dz, \tag{11.16}$$

where z_0 is the lowest point reached by the ray. The integrand contains an integrable singularity at $s = p$, the lowest point of the ray, where $\sin I = 1$. The data kernel is infinite at the bottom of the ray path, the turning point. No information can be retrieved from within low-velocity zones because no rays have turning points within low-velocity zones. The data kernel reflects this curious aspect of the problem: information is concentrated around the structure at the bottom of the ray path.

The methods of continuous inversion can be applied by changing the model to remove the singularity, a procedure known as *quelling*. Johnson and Gilbert (1972) do this by using integration by parts. In terms of the standard form (9.1)

$$\int_0^{z_0} K(x)s(x)dx = -\int_0^{z_0} Q(x)\frac{ds}{dx}dx, \tag{11.17}$$

where

$$Q(x) = \int_0^x K(x')dx'. \tag{11.18}$$

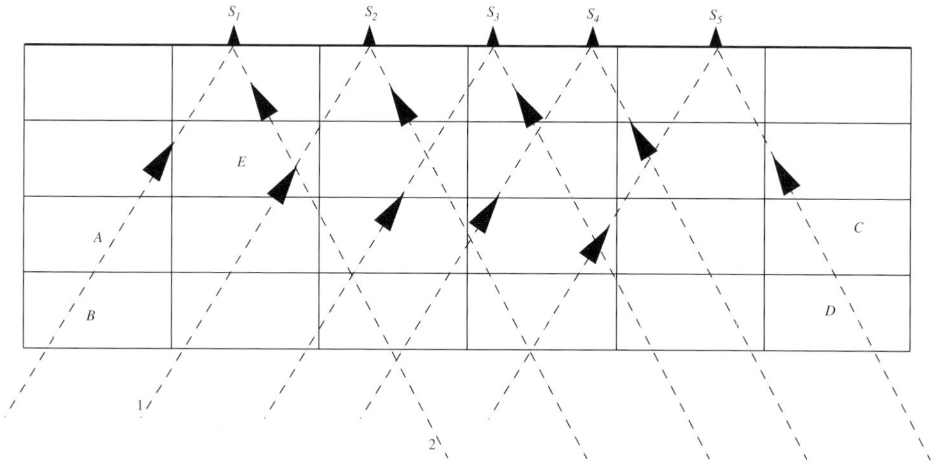

Fig. 11.5. Geometry of rays and receivers for the ACH inversion. Only one ray path passes through both blocks A and B, so there will be a trade-off between velocities in the two, creating an element of the null space. The same applies to blocks C and D. Ray paths 1 and 2 both pass through block E but have no others in common, which will allow separation of anomalies in block E from those in any other sampled block.

A final twist to the 1D problem comes from (Garmany, 1979), who showed it could be transformed into a linear inverse problem provided there are no low-velocity zones. Modern methods of inversion for one-dimensional structures are discussed in textbooks on seismology by Lay and Wallace (1995), Shearer (1999), and Gubbins (1990).

The inversion carried out by Aki *et al.* (1977) is a landmark in seismic tomography because of its simplicity of method and results. They determined relative arrival times from teleseisms (distant earthquakes) at the *NORSAR* array and inverted them for structure beneath the array. The assumptions seem severe at first sight: all ray paths from each event arrive parallel and can be treated as straight lines because of the great distance from the event, the mean arrival time from each event is unknown because of errors in location, all residual travel times are caused by anomalies immediately beneath the array to a fixed maximum depth comparable with the aperture of the array, and the structure can be represented by uniform rectangular blocks. The geometry of a typical ACH (Aki, Christoffersson and Husebye) inversion is shown in Figure 11.5.

The ACH problem is linear in the slowness $s(x, y, z)$ because we do no ray tracing beyond the initial paths. The parametrisation function $c_j(x, y, z)$ of equation (11.13) takes the values 1 inside the jth block and 0 outside it. The parameter m_j is the slowness anomaly in the jth block. The partial derivative $\partial T_i/\partial m_j$ is simply the length of the ith ray within the jth block. The mean travel time is subtracted for each event to remove uncertainties in travel time to the base of the assumed anomaly beneath the array caused by mislocation and unknown distant anomalies that affect all ray paths equally. This creates a null space containing the mean slowness in the layer because adding a constant slowness to all the blocks in any one horizontal layer will change the travel time of all the arrivals from one event by the same amount. Additional elements of the null space will include any blocks containing no ray paths, blocks where all ray paths are parallel (top row in Figure 11.2), and combinations of blocks where all rays have roughly equal length in both blocks.

In the simple ACH version of the method the null space can be readily understood and removed from the solution using the methods described in Chapter 7. Similar considerations indicate where we should expect resolution to be poor. Parallel rays inevitably occur near the surface (Figure 11.5), and the ACH method gives poor results for shallow structures, where by 'shallow' we mean a depth comparable with, or less than, the station spacing. The key to good resolution is to have crossing ray paths within the block. Resolution is therefore best at the base of the anomaly.

The literature contains a huge number of regional tomographic inversions following ACH principles. Most of them remove some assumptions made by ACH, such as straight rays and linearisation, although I remain unconvinced that these improvements change the results by much more than the errors inherent in the inversion. Ray tracing removes the null space because the rays are no longer parallel, but the geometry obviously leaves the average layer velocities very poorly determined and care is needed in interpreting the results. This should show up in the resolution matrix.

A common technique is to avoid the resolution matrix by a *checkerboard test*. A synthetic model consisting of alternating $+/-$ values is inserted into the matrix of blocks or model estimates, a synthetic set of travel times is calculated by ray tracing, and these are inverted in parallel with the real data. Recovery of the checkerboard pattern is used as good evidence of good resolution; places where the checkerboard squares are lost are viewed as evidence for poor resolution and poor results in that area.

In fact the checkerboard test does not evaluate the resolution properly: it merely tests one linear combination of the rows of the resolution matrix. To calculate the resolution matrix by using this sort of approach you would have to input a model with zero parameters everywhere except one; the result of the inversion would then be the row of the resolution matrix corresponding to the non-zero model parameter. In ACH inversion the worst-determined aspect of the model is certainly not the checkerboard, it is the mean layer velocity. Another simplified resolution test is to compute a spread function for each parameter (Michelini and McEvilly, 1991). This is useful, but again unsatisfactory, because the spread function computed is not the one optimised by the inversion and it contains only part of the whole story contained in the resolution matrix.

I see no need to avoid the resolution matrix. It is not that expensive to calculate and can be displayed, given some imagination, by using techniques for plotting functions of many variables: a matrix can be represented as a perspective plot as in Chapter 12 for example, and it looks nice in colour.

Modern developments in tomographic inversion include the use of many different seismic phases, including reflections and conversions from P to S waves at interfaces. Mao and Stuart (1997) developed a technique for exploration geophysics which combines transmission tomography with reflection data. Eberhart-Phillips and Reyners (1999) have used converted phases from the subducted Pacific plate beneath the North Island of New Zealand in a tomographic inversion that reveals the depth of the subducted plate rather well because the converted phases resolve the depth well. Large-scale tomographic inversions, such as those by Grand (1987) for North America and van der Hilst *et al.* (1991) for subduction zones around the world have made an enormous impact on our understanding of structures in the lithosphere and mantle.

11.3 Simultaneous inversion for structure and earthquake location

A very common situation is to invert arrival times for structure when the source locations are uncertain. For example, a temporary deployment of seismometers may record a large number of local and regional earthquakes in an active tectonic area. The local structure is inhomogeneous because of the active tectonics, and this produces differences from arrival times calculated from a standard velocity model. Uncertainties in the arrival times frustrate efforts to locate earthquakes accurately, and earthquake mislocation frustrates efforts to determine the structure. The proper formal procedure is to invert for both hypocentral coordinates and structure simultaneously, and this is now often done.

The combined inverse problem exhibits some general characteristics of mixing two different types of model parameters, in this case the positions of the hypocentres and slowness parameters. It is a combination of earthquake location, discussed in Chapter 8, and tomography, discussed in the previous section. The travel-time integral is taken between the unknown source position (x_H, y_H, z_H) and the known station position S,

$$T_{HS} = \int_{H(x_H, y_H, z_H)}^{S} s(x, y, z)\mathrm{d}l, \tag{11.19}$$

where the integral is taken along the ray path. The inverse problem is linearised as before, using Fermat's principle. The partial derivatives with respect to the hypocentre parameters is obtained by differentiating the integral with respect to its lower limit. The results are similar to those obtained in Chapter 8, with the azimuth and incidence angle being determined from the take-off angles of the ray at the source location. In a general 3D structure we cannot assume the ray path remains in the plane of the source and receiver and the azimuth must be calculated from the ray path rather than from the source–receiver geometry.

The slowness is parametrised, although much of the following discussion also applies to the continuous inverse problem. The linearised problem then has the joint form

$$\delta \boldsymbol{T} = \mathsf{B}\boldsymbol{h} + \mathsf{A}\boldsymbol{m} + \boldsymbol{e}, \tag{11.20}$$

where B is the usual matrix of partial derivatives with respect to the hypocentral coordinates and $\boldsymbol{h}$ is the four-vector of small improvements to the hypocentral coordinates, A is the matrix of partial derivatives with respect to the slowness parameters, $\boldsymbol{m}$ is the four-vector of small improvements to the velocity model, computed by integrating the basis functions for the parametrisation along the existing known ray paths, as in (11.14).

Now $\boldsymbol{h}$ and $\boldsymbol{m}$ are combined into a single model vector, B and A are combined into a single equation of condition matrix, and (11.20) is written in partitioned form

$$\delta \boldsymbol{T} = (\mathsf{B}|\mathsf{A}) \left(\frac{\boldsymbol{h}}{\boldsymbol{m}} \right) + \boldsymbol{e}. \tag{11.21}$$

We need to scale the model parameters to approximately the same size: a good rule of thumb is to reduce them to the same dimensions. The idea is to produce model parameters scaled in such a way that similar changes in each of them produce similar changes in the arrival time. For example, a change in origin time of 1 s will produce a change in arrival time of 1 s; a shift in hypocentre of 1 km (along the ray path) produces a change in arrival time of v_H^{-1} seconds, where v_H is the seismic velocity at the source location. v_H is therefore a suitable scale for the depth and epicentral parameters. Similarly, multiplying the slowness parameters by a typical ray path length gives a typical change in arrival time from changes in the structure. We must not use any consideration of data error in these scalings; they are irrelevant to our choice of model. Once scaled, we can use a single form of damping for all the scaled parameters the same way, should damping be needed.

Forming a single, standard inverse problem has some mathematical appeal but it carries several serious disadvantages. It conceals the underlying structure of the problem, the essential differences between a source location and velocity; it makes it difficult to assess the distribution of uncertainties between mislocation and velocity determination, and it is computationally inefficient unless some clever algorithm is used to take advantage of the sparse nature of the combined equations of condition matrix.

Equation (11.20) makes clear the joint nature of the inversion. If $\boldsymbol{h} = 0$ we have tomography, if $\boldsymbol{m} = 0$ we have earthquake location. If we can find some way to transform the equations (which means transforming the data) to eliminate either term we could begin to discuss the solution in terms of one or other of the simpler subproblems. Minimising the sum of squares of errors $\boldsymbol{e}^{\mathrm{T}}\boldsymbol{e}$ with respect first to $\boldsymbol{h}$ then to $\boldsymbol{m}$ gives the two sets of normal equations

$$\mathsf{A}^{\mathrm{T}}\delta\boldsymbol{T} = \mathsf{A}^{\mathrm{T}}\mathsf{A}\boldsymbol{m} + \mathsf{A}^{\mathrm{T}}\mathsf{B}\boldsymbol{h} \tag{11.22}$$

$$\mathsf{B}^{\mathrm{T}}\delta\boldsymbol{T} = \mathsf{B}^{\mathrm{T}}\mathsf{A}\boldsymbol{m} + \mathsf{B}^{\mathrm{T}}\mathsf{B}\boldsymbol{h}. \tag{11.23}$$

Bearing in mind that only matrices $\mathsf{A}^{\mathrm{T}}\mathsf{A}$ and $\mathsf{B}^{\mathrm{T}}\mathsf{B}$ are square and invertible, we use (11.23) to eliminate $\boldsymbol{h}$ from (11.22) in favour of $\boldsymbol{m}$ to give

$$\boldsymbol{m} = (\mathsf{OA})^{-1}\mathsf{O}\delta\boldsymbol{T} \tag{11.24}$$

$$\boldsymbol{h} = (\mathsf{B}^{\mathrm{T}}\mathsf{B})^{-1}[\mathsf{B}^{\mathrm{T}}\delta\boldsymbol{T} - \mathsf{B}^{\mathrm{T}}\mathsf{A}\boldsymbol{m}], \tag{11.25}$$

where

$$\mathsf{O} = \mathsf{A}^{\mathrm{T}} - \mathsf{A}^{\mathrm{T}}\mathsf{B}(\mathsf{B}^{\mathrm{T}}\mathsf{B})^{-1}\mathsf{B}^{\mathrm{T}}. \tag{11.26}$$

This formulation is computationally very efficient because the matrix $\mathsf{B}^{\mathrm{T}}\mathsf{B}$ comprises 4×4 blocks along the diagonal; all other elements are zero. B is the partial derivatives matrix with respect to the hypocentral parameters, each arrival time only depends on one source location, so each row of the matrix B has only four non-zero entries. Inversion of $\mathsf{B}^{\mathrm{T}}\mathsf{B}$ involves n_e inversions of 4×4 matrices, where n_e is the number of sources. This is a much faster task than inverting a single $4n_e \times 4n_e$ matrix.

Matrix O is an operator that annihilates $\boldsymbol{h}$ from equation (11.20). It is easy to show that $\mathsf{OB} = 0$, so premultiplying (11.20) by O gives

$$\mathsf{O}\delta\boldsymbol{T} = \mathsf{OA}\boldsymbol{m} + \mathsf{O}\boldsymbol{e}. \tag{11.27}$$

Transforming the data by premultiplying by O leaves a purely tomographic problem; the transformed errors $\mathsf{O}\boldsymbol{e}$ carry information about the effects of mislocation on the error analysis for the tomography.

Returning to equations (11.22) and (11.23), we could alternatively use (11.22) to eliminate $\boldsymbol{m}$ in favour of $\boldsymbol{h}$ in (11.23). This gives the alternative solution

$$\boldsymbol{h} = (\mathsf{QB})^{-1}\mathsf{Q}\delta\boldsymbol{T} \tag{11.28}$$

$$\boldsymbol{m} = (\mathsf{A}^{\mathrm{T}}\mathsf{A})^{-1}[\mathsf{A}^{\mathrm{T}}\delta\boldsymbol{T} - \mathsf{A}^{\mathrm{T}}\mathsf{B}\boldsymbol{h}], \tag{11.29}$$

where

$$\mathsf{Q} = \mathsf{B}^{\mathrm{T}} - \mathsf{B}^{\mathrm{T}}\mathsf{A}(\mathsf{A}^{\mathrm{T}}\mathsf{A})^{-1}\mathsf{A}. \tag{11.30}$$

This time Q annihilates $\boldsymbol{m}$ from (11.20) leaving the earthquake location problem

$$\mathsf{Q}\delta\boldsymbol{T} = \mathsf{QB}\boldsymbol{h} + \mathsf{Q}\boldsymbol{e}. \tag{11.31}$$

Assessing contributions to model covariances and resolution from different sets of model parameters is difficult. This problem is discussed further by Spencer (1985) and the proper treatment of resolution matrices is dealt with fully in the book by Ben-Israel and Greville (1974).

Several methods have been devised for accurate relative locations of earthquakes. These include joint hypocentre determination (JHD) (Douglas (1967); Pavlis and Booker (2000); Pujol (1988)) and the related master-event methods, the homogeneous station method (Ansell and Smith (1975)), and the recent double-differencing method (Waldhauser and Ellsworth (2000)).

In JHD one takes a cluster of earthquakes and assumes that they are sufficiently close together that the ray paths to any one station all pass through the same anomalies and that the arrival times are all affected by the same time error. This error is removed by using differences of arrival times rather than the times themselves. The events can be accurately located relative to each other because large errors arising from uncertainties in structure are removed. The absolute location is not well determined. In terms of our formalism, taking differences of arrival times at stations is the operation performed by premultiplying by Q to form (11.31); matrix QB then has a null space corresponding to the mean location of the cluster of earthquakes.

In the master-event method it is assumed that one event, the master, is well located. This is often the case when a large earthquake is followed by a number of smaller aftershocks; the large event is recorded by a greater number of stations and is therefore better located than the aftershocks. The master-event method is similar to JHD, but the one (or possibly more) well-located event combined with good relative locations of the smaller events to the master produce good absolute locations for all events. The same effect can be produced by using an explosion with a precisely known location in the source region.

For example, consider equation (11.20) written out explicitly for the ith event and jth station, and subtract the times for the master event $i = 0$:

$$\delta T_{ij} - \delta T_{0j} = \sum_{k=1}^{4} B_{ij,k} h_k + \sum_{r-1}^{P} (A_{ij,k} - A_{0j,k}) m_r + (e_{ij} - e_{0j}). \tag{11.32}$$

The master event is supposed to be well located so no improvement is needed to its hypocentral coordinates. If the ray paths are sufficiently similar for the two events, the differences in partial derivatives $(A_{ij,k} - A_{0j,k})$ will be small and the effects of velocity structure is minimised.

In the homogeneous station method the same principle is applied by differencing arrival times at stations rather than differencing times from each event.

12

Geomagnetism

12.1 Introduction

Global mapping of the Earth's main magnetic field is a very old scientific problem, dating from the earliest times of navigation in the days of European exploration. Measurements of the magnetic field are used to construct a global map of all three of its components. Initially measurements of one component, the declination or angle between the compass direction and true North, were hand-contoured on a map. The modern era dates from C. F. Gauss, who developed the theory for representing the magnetic field in terms of spherical harmonics (Section 12.2), least-squares analysis, and an early network of magnetic observatories. Since the 1960s a reference magnetic field, the International Geomagnetic Reference Field (IGRF), has been published every five years and, by international agreement, is used as the standard for mapping purposes. The IGRF is now built into GPS instruments to convert between true and magnetic bearings, and is used to define the compass direction on maps and marine charts. Updating is needed every five years because the Earth's magnetic field changes by a per cent or two during that time. The IGRF poses a classic parameter estimation problem: we need a concise, accurate model of the magnetic field using a simple representation with as few parameters as possible. It is not well suited to scientific study of the magnetic field because of the nature of international agreement: the criteria vary from epoch to epoch depending on the available data and even on political considerations from the contributing countries.

I have been involved with the problem of constructing a model of the geomagnetic field for scientific purposes, and describe here some of the methods we devised to obtain a consistent model of the historical period (from AD 1550 to the present day) for scientific purposes. The data have varied greatly in both quality and type over the centuries (Section 12.5), making it impossible to obtain a model with uniform accuracy in time. It is possible to construct a model with error bounds consistent with basic assumptions about the nature of the geomagnetic field (Section 12.3), the most important being its origin in the Earth's core 2886 km below the Earth's surface. Inverse theory is ideally suited to this problem. Construction of the magnetic field at depth within the Earth, done by a process known as downward continuation, is unstable. This means the errors amplify with increasing depth, making it very difficult (some might say impossible) to determine the magnetic field at the surface of the Earth's core. For the same reason, a prior requirement that the magnetic field be reasonably smooth at the core surface is a very strong constraint on the model for the magnetic field at the Earth's surface (Section 12.4) because the errors associated with small wavelengths must not allow the model to grow drastically when downward continued.

The main purpose of developing these geomagnetic models was to understand the processes underway in the Earth's liquid core. Changes in the geomagnetic field are caused by fluid flow at or near the core surface, and they can be mapped into core flow provided electrical resistance can be neglected. Under these conditions the magnetic field is frozen to the fluid and acts as a tracer for the flow. Unfortunately, it is not possible to label individual magnetic field lines, and although we can follow some magnetic features in time it is not possible to observe the fluid flow directly: the inverse problem of finding fluid flow from the secular variation or rate of change of the magnetic field has a null space. Much recent work has been directed towards reducing this null space by making additional assumptions about the flow based on the fluid dynamics operating at the top of the core. This produces a fascinating example of an ill-posed inverse problem (Section 12.8).

12.2 The forward problem

The Earth's atmosphere, crust, and mantle may be treated as electrical insulators on the long periods, longer than about a year, that are of interest in mapping the main field. There are no electric currents and the magnetic field $\boldsymbol{B}$ may be written as the gradient of a magnetic potential V,

$$\boldsymbol{B} = \nabla V, \tag{12.1}$$

which satisfies Laplace's equation

$$\nabla \cdot \boldsymbol{B} = \nabla^2 V = 0. \tag{12.2}$$

Laplace's equation may be solved everywhere in the insulator provided we have satisfactory boundary conditions, such as V, $\boldsymbol{B} \cdot \boldsymbol{n}$ where $\boldsymbol{n}$ is normal to the surface, or a linear combination of the two.

A general solution of Laplace's equation is available in spherical coordinates (r, θ, ϕ) as an expansion in spherical harmonic functions. In geomagnetism this is usually written as

$$V(r, \theta, \phi) = a \sum_{l,m} \left(\frac{a}{r}\right)^{l+1} P_l^m(\cos\theta)\left(g_l^m \cos m\phi + h_l^m \sin m\phi\right) \tag{12.3}$$

(see, for example, Jacobs (1994); Arfken and Weber (2001); Kellogg (1953); Backus *et al.* (1996)). Here a is the Earth's radius, P_l^m is an associated Legendre function, and g_l^m, h_l^m are the *geomagnetic coefficients* describing the magnetic field, sometimes called the Gauss coefficients. $P_l^m(\cos\theta)\cos m\phi$ and $P_l^m(\cos\theta)\sin m\phi$ are called *spherical harmonics*. Functions with large l are small scale. Expansion in spherical harmonics is analogous to Fourier series on spheres and there are many results parallel with those given in Part I of this book.

Equation (12.3) is used to derive the forward problem, the formula that allows us to calculate the magnetic field components. For example, the vertical component Z measured at the Earth's surface (by convention Z is measured positive downwards) is obtained by differentiating (12.3) with respect to r and setting $r = a$

$$Z(a, \theta_0, \phi_0) = \sum_{l,m}(l+1) P_l^m(\cos\theta_0)\left(g_l^m \cos m\phi_0 + h_l^m \sin m\phi_0\right). \tag{12.4}$$

The nature of measurements has changed over the years with the development of instrumentation. We never measure magnetic potential directly. The first, and easiest, measurement used to be the declination, D, which was the only measurement available until the late sixteenth century. Inclination, I, followed, and Gauss developed the first method of measuring absolute intensity or field strength early in the nineteenth century. Seven different components have been measured in all. These provide the basic data for the inversion: the three spherical components X, Y, Z; the angles D, I; and the total and horizontal intensities F, H. Of these, X, Y, Z are linearly related to the geomagnetic coefficients (e.g. equation (12.4)): inverting them poses a linear inverse problem. All the others are nonlinearly related and pose a nonlinear inverse problem. For example, the declination involves the arctangent:

$$D = \tan^{-1}\left(\frac{Y}{X}\right). \tag{12.5}$$

12.3 The inverse problem: uniqueness

The model vector is defined as simply the array of geomagnetic coefficients arranged in a suitable order

$$\boldsymbol{m} = \left(g_1^0, g_1^1, h_1^1, g_2^0, g_2^1, h_2^1, \ldots\right). \tag{12.6}$$

The data kernels are the derivatives of the spherical harmonics at the measurement site. For example, (12.4) shows that the data kernel for a measurement of the vertical component is $(l+1)P_l^m(\cos\theta_0)\cos m\phi_0$ for a g coefficient ($\sin m\phi_0$ for an h coefficient). For the chosen range of l and m the data kernels form the row of the equation of condition matrix corresponding to this particular vertical component measurement at a particular site.

Nonlinear data, such as D, are incorporated by a quasi-linearised iteration as described in Chapter 8. The data kernels are obtained by differentiation. Consider horizontal intensity H first. From the definition

$$H = \sqrt{X^2 + Y^2} \tag{12.7}$$

differentiation gives a small change in H, δH, in terms of small changes in X and Y

$$\delta H = \frac{1}{H}(X\delta X + Y\delta Y). \tag{12.8}$$

Since X and Y are linearly related to the model, (12.8) gives a linear relationship between a small change in horizontal intensity and small changes in the geomagnetic coefficients. These are the quantities needed for the row of the equations of condition matrix when doing an iterative inversion. Similarly, differentiating (12.5) gives the required formula for a measurement of declination

$$\delta D = \frac{1}{H^2}(X\delta Y - Y\delta X). \tag{12.9}$$

In each case values of the components D, H, F, etc., are computed from the model of the previous iteration.

Potential theory is well understood and standard results give us much of the information we need about uniqueness of the inversion given perfect data. For example, a standard theorem guarantees uniqueness if the normal gradient of the potential is known everywhere on the bounding surface of the region where we wish to solve for the potential. Thus knowing the vertical component everywhere on the Earth's surface would produce a unique solution. The North component, B_θ in spherical coordinates, also gives a unique solution, but the East–West, or azimuthal component, does not. This is easily shown by calculating the data kernel for B_ϕ by differentiating (12.3) with respect to ϕ. Clearly all terms involving $m = 0$ vanish on differentiation; the equations of condition do not involve any geomagnetic coefficients with $m = 0$, so this part of the main field remains completely undetermined. So $m = 0$ defines the axisymmetric part of the magnetic field, and the axisymmetric part forms the null space for an inverse problem in which only East–West components of field are measured.

Datasets with only East–West components almost never arise, but satellite data are often confined to total intensity data, and such data often dominate in the modern era. Backus (1970a) showed that total intensity measurements do not provide a unique solution, but leave a null space now known as the Backus ambiguity. It is characterised by a spherical harmonic series dominated by terms with $l = m$. The Backus ambiguity is alleviated by even a modest amount of ground-based measurements of other components.

Prior to 1839 we had no absolute intensity measurements, only directions. It is then obvious that any field model could be multiplied by an arbitrary scalar constant and still fit the data. Interestingly, the null space for a linearised, iterative solution comprises any constant multiplied by the best-fitting model itself, since addition of any multiple of the model simply multiplies the model by a constant factor, which does not alter the fit to the data. It was originally assumed implicitly that a single constant was the full extent of the ambiguity, but Proctor and Gubbins (1990) showed for axisymmetric fields that more unknown constants could be involved, and conjectured a similar general result. The result in 3D was finally proved by Khokhlov *et al.* (2001). The full solution to the inverse problem for perfect directional data covering the Earth's surface involves a linear combination of n fields, with n unknown coefficients, where n is one less than the number of dip poles. A strange result, but a reassuring one when one considers that the Earth's magnetic field has only two dip poles, and therefore the original implicit assumption of just one arbitrary constant was correct.

Finally, in the earliest times datasets were dominated by declination. Gubbins (1986) showed that perfect declination data combined with measurements of horizontal intensity along a line joining the dip poles are sufficient to provide a unique solution, another strange result. Measurement of inclination on a voyage from France around South America into the Pacific, and another from England to the East Indies, were enough to provide very good control on a global model for epoch AD 1695–1735, which includes a large volume of declination data (Bloxham, 1986). In practice, data coverage of the globe is sparse and the data contain errors, but formal uniqueness theorems provide an important guide to the type of data that are needed to control the solution well.

Proctor and Gubbins's (1990) study gives an example of how great can be the difference between the original, continuous problem and the discretised version. They computed synthetic data for an axisymmetric field with two lines of dip poles, then inverted by standard methods using a truncated spherical harmonic series like (12.3). They found both solutions, but only after truncating the series well above degree 50. This is much higher than one would ever use for the Earth's main field (nowadays the IGRF is trunctated at degree 12) and higher than would normally be needed for a satisfactory numerical representation of either solution of this particular synthetic problem. The high truncation was needed to obtain the second family of solutions: truncating below degree 50 yielded only one solution. This is an example of the danger of extrapolating uniqueness of a discretised problem to the continuous case, and underlines the value of assessing the fundamental properties of the inverse problem at the outset.

12.4 Damping

Now we must distinguish between mapping the magnetic field at the Earth's surface, the purpose of the IGRF, and developing magnetic field models for scientific purposes. In the former case we seek only a concise and accurate model of the magnetic field components near the measurement sites. The usual procedure is to truncate the expansion (12.3) at some suitable degree and apply the methods of Chapter 6 to find estimates of the geomagnetic coefficients. The truncation point has increased over the years from 8 to 12; its selection is based on the degree where the signal from the crust, with its origin in magnetised surface rocks, begins to dominate the magnetic spectrum[1]. Since one purpose of the IGRF is to subtract a well-defined large-scale field from small-scale regional survey maps, this choice of truncation makes considerable sense.

In scientific studies we need to quantify how accurately we separate the core field, and the parameter estimation approach is unsuitable. The model is really that part of the magnetic field that originates in the core; the signal from the Earth's crust is just another source of noise. The normal component of magnetic field on any surface provides a unique solution to Laplace's equation and therefore defines the magnetic field everywhere within the insulating region. We can therefore take the model to be the radial component of magnetic field on the core surface and formulate a continuous inverse problem.

The magnetic field at the Earth's surface ($r = a$) is related to the model by a surface integral (Gubbins and Roberts (1983))

$$\boldsymbol{B}(a, \theta, \phi) = \oint B_{\mathrm{r}}(c, \theta', \phi')\boldsymbol{N}(\mu)\mathrm{d}S, \tag{12.10}$$

where the integral is taken over the core surface and $\mu = \cos\alpha$ is the cosine of the angular distance between the two points (θ, ϕ) and (θ', ϕ'). The components of the vector $\boldsymbol{N}$ are the data kernels that describe how a particular measurement at the

[1] On a sphere the spherical harmonic coefficients are correctly referred to as the spectrum, with degree l taking the part of the wavenumber.

Earth's surface samples the vertical component at the core surface. They are given by formulae involving the generating function for associated Legendre functions:

$$N_r(\theta, \phi; \theta', \phi') = \frac{c^2}{4\pi a^2} \frac{(1 - c^2/a^2)}{f^3} \tag{12.11}$$

$$N_\theta(\theta, \phi; \theta', \phi') = -N_h[\cos\theta \cos\theta' \cos(\phi - \phi') - \sin\theta \cos\theta'] \tag{12.12}$$

$$N_\phi(\theta, \phi; \theta', \phi') = -N_h \sin\theta' \sin(\phi - \phi') \tag{12.13}$$

$$N_h(\theta, \phi; \theta', \phi') = \frac{c}{4\pi a}\left[\frac{1}{1+\mu} + \frac{c/a - f}{f(f - c/a + \mu)} - \frac{2c^2}{f^3 a^2}\right] \tag{12.14}$$

$$f(r; \theta, \phi; \theta', \phi') = \frac{1}{a}(a^2 - 2ac\mu + c^2)^{\frac{1}{2}}. \tag{12.15}$$

$$\tag{12.16}$$

The functions N_r and N_h depend only on the angular distance between the measurement point on the Earth's surface and the sample point on the core–mantle boundary. They are plotted in Figure 12.1. N_r peaks at $\alpha = 0$, showing that a vertical field measurement at the Earth's surface best samples the vertical field at the core surface immediately below the measurement site. It also samples a considerable area of the core surface around the point directly below the

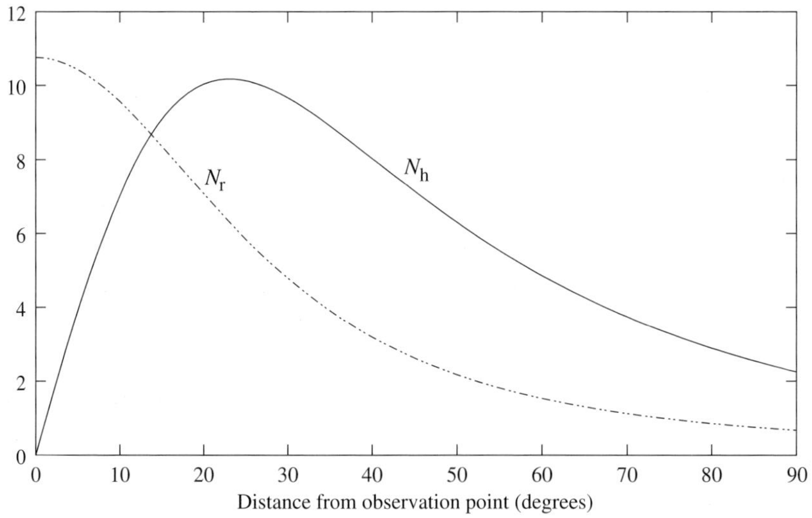

Fig. 12.1. Data kernels for the continuous inverse problem of finding the radial component of magnetic field on the core surface from a measurement of radial component (N_r) and horizontal component (N_h) at the Earth's surface as a function of angular distance between the two points. Note that radial component measurements sample best immediately beneath the site, but horizontal components sample best some 23° away.

measurement point, the data kernel only falling to about half its peak value at a distance of 28°. N_{h} shows how horizontal components sample the core surface. The response is zero for $\alpha = 0$, showing that a surface measurement of the horizontal field does not sample the vertical field immediately below it. The data kernel increases to a maximum at a distance of 23°. The angle between the vertical on the core surface and the horizontal on the Earth's surface is $90° - \alpha$; the kernel initially increases with α because this angle decreases, introducing a component of core field in the measured direction. It ultimately decreases because of the greater distance between the core field and measurement point. This means horizontal components sample the core field a considerable (angular) distance from the measurement site: a simple map of observation points is therefore not a good indicator of geographical coverage. The remaining contributions to N_θ and N_ϕ in (12.12) and (12.13) contain the variation with azimuth around the measurement site: N_ϕ is more sensitive to core fields to the east and west; N_θ to the North and South.

The continuous formulation has never been implemented; everyone has preferred the relative simplicity of working from the spherical harmonic expansion (12.3). The following inversion is from Bloxham *et al.* (1989), which follows the general procedure outlined in Chapter 7. We truncate (12.3) at some high level, retaining many coefficients beyond those that could be estimated accurately. We apply a prior constraint on the magnetic field on the core surface.

Minimising the integral of the square of the magnetic field on the core surface is one choice

$$\oint B_{\mathrm{r}}^2 \mathrm{d}S = C \sum_{l,m} (l+1) \left(\frac{a}{c}\right)^{2l} \left(g_l^{m2} + h_l^{m2}\right) = \theta^2 \boldsymbol{m}^{\mathrm{T}} \mathsf{W} \boldsymbol{m}, \tag{12.17}$$

where C is a known constant and the weight matrix W is diagonal. Consider the diagonal elements of W. As l increases the element increases by a factor $(l+2)/(l+1)$ times $(a/c)^2 = 3.342$. The elements of the damping matrix increase as l increases, so the shorter wavelengths are damped more heavily than the longer wavelengths. This is exactly what we want in order to produce a regular solution to the inversion because the short wavelengths are less well constrained by the data than the long wavelengths. When l is large, above 10 say, the first factor $(l+2)/(l+1)$ approaches unity but the downward continuation factor $(a/c)^2$ remains the same whatever the value of l. This factor provides by far the strongest part of the damping.

The electrical heating is a better physical constraint than the surface energy because we know its upper limit from estimates of the heat passing across the core–mantle boundary. It is also a stronger constraint because it involves the square of the electric current and therefore the curl of the magnetic field. It is not possible to compute the electrical heating from the geomagnetic coefficients because we cannot continue the solution (12.3) inside the core. Gubbins (1975) derived an inequality for the electrical heating in terms of the geomagnetic coefficients by choosing an optimal form for the magnetic field within the core that minimised the electrical heating, but the inequality is so far from any likely equality that the result is not useful. Instead, most authors have used rather arbitrary damping where W is taken to be diagonal with elements $l^n(a/c)^{2l}$ where n takes values from 1 to 6. The damping has a significant effect on the solution only when l is large, so the $(l + 1)$ can be replaced by l in (12.15). At high values of l the data have no effect on the solution and everything depends on the choice of damping. This allowed Bloxham *et al.* (1989) to calculate the effects of raising the truncation of the spherical harmonic series to arbitrarily high levels: errors introduced by the truncation were thereby eliminated to give the same solution as a continuous inversion with the same prior assumptions. In practice they found the asymptotic limit was reached by truncating at degree 20.

Core-surface damping produces a smooth field model, free of any ringing effects associated with premature truncation of the spherical harmonic series. It provides powerful prior information for a Bayesian inversion, and should give very reliable error estimates for the Earth's surface field.

The same remarks do not apply to maps of the magnetic field at the core surface. Downward continuation to the core amplifies the geomagnetic coefficients by the factor $(a/c)^l$ and the errors are amplified by the same amount. The errors on the core surface are therefore controlled only by the additional assumptions in the prior model covariance matrix, the rather arbitrary factor l^n and the damping constant θ^2. Quoting errors for field models on the core surface is controversial (Backus, 1988).

12.5 The data

The historical dataset is very inhomogeneous because of its long, 400-year time span. Different instruments have been developed during the period and magnetic surveying has taken place in spurts, which have caused dramatic changes in the geographical coverage over time. Early datasets are dominated by declination, with only a handful of inclination measurements available for the seventeenth century. Feuillée's voyages, starting in 1714, were the first to provide anything like global coverage of inclination. These early voyages suffered from poor navigation; uncertainty in longitude on sea voyages was a major contributor to error until Cook's expeditions starting in 1775. Errors in position produce uncertainties in the equations-of-condition matrix A. We have not dealt with this type of error in this book; the reader is referred to Jackson *et al.* (2000) for a recent discussion in the context of this problem. Intensity data became widely available from 1840.

Table 12.1. *Some key historical dates in the development of magnetic surveying*

2000	Oersted satellite heralds decade of magnetic observation
1980	Magsat – first vector measurements by satellite
1960	POGO and other scalar measurements by satellite
1955	Proton magnetometer makes magnetic surveys easy, particularly at sea
1926	Research vessel *Carnegie* burns in Apia harbour causing gap in marine surveys
1880	Challenger expedition provides global coverage
1840	Royal Navy survey of southern oceans, Ross expedition to Antarctica
1830	Gauss develops method of measuring absolute intensity
1775	Cook's voyages, navigation much improved
1700	Halley's cruises of the Atlantic in the *Paramour*
1601	William Gilbert publishes *De Magnete*
1586	Robert Norman invents the inclinometer

The early twentieth century saw the start of detailed land surveys (prompted by mineral exploration) and the establishment of permanent magnetic observatories capable of monitoring the secular variation. Later datasets became dominated by total intensity because of the convenience of the proton magnetometer. This continued into the satellite era because of the ease of measuring total intensity compared with vector components, which requires star cameras to determine orientation. We are now in a period when many ground-based observatories are closing but several new satellite missions are starting. Some historical landmarks are listed in Table 12.1 and data distributions are plotted in Figure 12.2.

In all but the earliest measurements data errors tend to be dominated by signal from magnetised rocks in the crust. This contributes an average of 200–300 nT and a maximum, over seamounts and volcanic terrain, of over 1000 nT. In terms of direction these errors translate to 0.1–3° over distances of several hundred kilometres. Directions must be weighted to convert them to an intensity in order to provide a uniform dataset for inversion (this could equally be considered as part of the data covariance matrix C_e). D must be weighted with H because the accuracy of a directional measurement is determined by the force acting on the compass needle, which is proportional to H (note that declination is not determined at all at the geomagnetic poles, where $H = 0$). Similarly, I must be weighted with F because the accuracy of a measurement of I is determined by the force exerted on an inclinometer needle, which is proportional to F.

Intensities themselves were given a nominal error of 300 nT for F and 200 nT for all single components including weighted directions by Bloxham *et al.* (1989) in order to account for the crustal 'noise' as well as instrumental error. This gives a diagonal covariance matrix; the values can be justified to some extent by the final fit of the models to the weighted data, which are quite close to 1. These

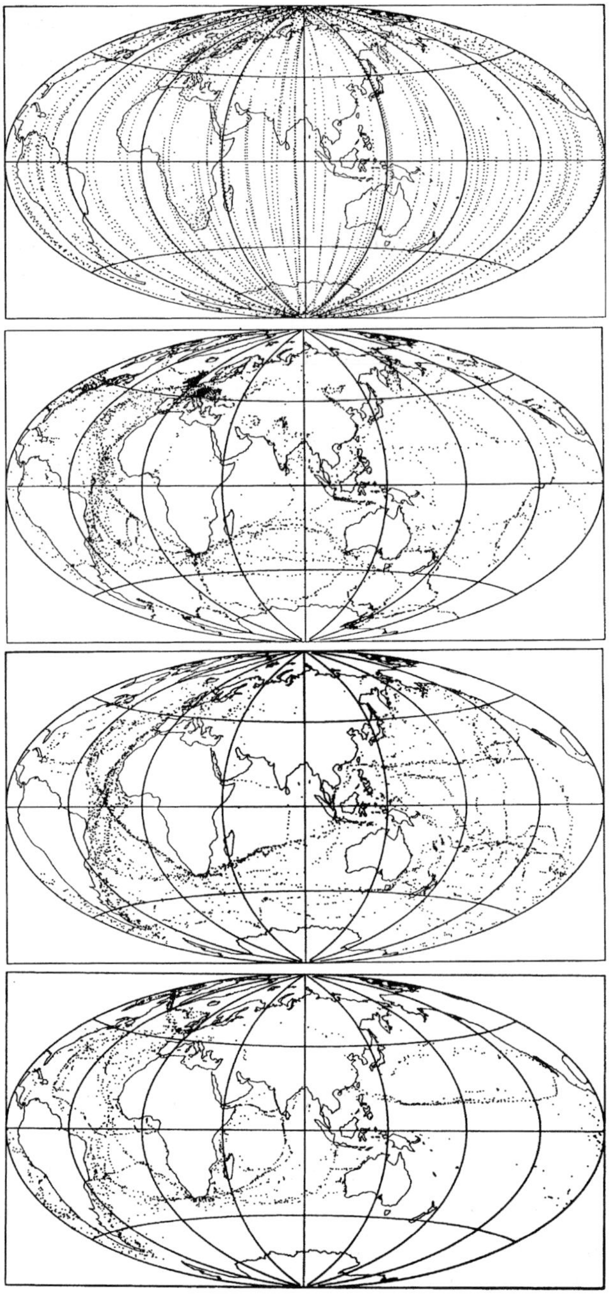

Fig. 12.2. Data distribution plots. From the top: epoch 1966, dominated by total intensity from the POGO satellite; epoch 1842, bearing Gauss's influence on magnetic observation; epoch 1777.5, the time of Cook's voyages; and epoch 1715, containing Halley's voyages and Feuillée's measurements of inclination. Reproduced, with permission, from Bloxham *et al.* (1989).

errors apply only at ground level. At satellite altitude the crustal signal is much smaller because its small length scale is attenuated strongly with height. A smaller error needs to be applied; 10 nT is typical for Magsat altitude. External magnetic fields are another source of error. Ground-based measurements can be assumed to have been made at magnetically quiet times, but satellites fly through polar regions where charged particles flow along magnetic field lines and create field-aligned currents. These disturb the horizontal components of magnetic field more than the vertical components. Only vertical component data from Magsat were used above 50° magnetic latitude.

Permanent magnetic observatories monitor the time variation of the magnetic field over decades and it is possible to remove the crustal signal in the form of a station correction. The nominal accuracy of absolute measurements made at these observatories is 5 nT, but this is very difficult to achieve in practice. Contributions from atmospheric disturbances are hard to eliminate completely, and maintaining the standard over decades is virtually impossible because of changes in instrumentation and staff. Many 'permanent' observatories have been forced to change location because of encroaching urban development. Efforts are often made to link the measurements at the new site to those at the old site but these are rarely fully satisfactory. Errors can be estimated by plotting the data as a function of time and estimating the scatter from a straight line or smooth curve. An example was shown in Figure 6.3.

Bloxham *et al.* (1989) binned the data into decade intervals for the twentieth century and into larger time bins for earlier centuries. The number and type of data are shown in Table 12.2. Outliers were removed progressively from the inversions by simple data rejection; numbers of data rejected at the 3σ standard deviation level in the final iteration are listed in the table.

12.6 Solutions along the trade-off curve

The general form of the damping is specified by the matrix W; the precise damping requires choice of the constant θ^2. Four rationales were given in Chapter 7: (a) to fit the data (misfit to the weighted data equal to one), (b) to produce a uniform model (norm equal to some prescribed value), (c) to incorporate prior information in a Bayesian inversion (the right Ohmic heating for example), and (d) to find the 'knee' of the trade-off curve.

In practice we compute a series of models and construct a trade-off curve. From this we can read off the value of the damping constant for unit misfit, a given norm, or the knee. The series was truncated at spherical harmonic degree 20, giving 440 parameters to be determined. Table 12.3 gives numerical details of each model. The trade-off curve for epoch 1980 is shown in Figure 12.3. The scales are

Table 12.2. *Details of the data contributing to each model. The model year is the epoch to which all data were reduced (except the two oldest models, for which no secular variation estimates were available). The median year is the average date of all the contributing survey data (excluding the observatory data). The remaining columns give the number of measurements by component; the differences in their distribution between epochs is mainly the result of the availability of different instruments down through the ages*

Model year	Median year	Data total	Data rejected at 3σ	D	H	I	X	Y	Z	F
1980.0	1980.0	4178	0	0	0	0	1262	1262	1654	0
1966.0	1966.0	11040	21	424	392	117	146	146	460	9334
1955.5	1957.4	30732	319	9335	4168	6754	136	135	2963	6922
1945.5	1945.5	16208	367	7079	4290	3135	86	86	1084	81
1935.5	1936.1	16700	506	8154	4089	2477	96	96	1282	0
1925.5	1924.4	23210	554	10711	6028	5567	62	62	225	0
1915.5	1914.8	24865	393	9812	7489	7040	51	51	56	0
1905.5	1905.4	20203	267	7639	6029	6106	47	47	47	0
1882.5	1880.0	10573	534	5214	1529	2601	39	39	37	585
1842.5	1844.2	12398	380	4335	0	4615	0	0	0	3068
1777.5	1776.9	6115	288	4551	0	1276	0	0	0	0
1715.0	1715.7	2636	81	2508	0	128	0	0	0	0

Table 12.3. *Model parameters. The models 1905.5–55.5 were derived from data that were weighted differently from the other models and the misfits may be adjusted upwards accordingly*

Model year	Damping constant/10^{-14}	Norm /10^{12}	Misfit	Resolution
1980.0	1000	136	1.09	150
1966.0	2000	130	1.29	144
1955.5	200	132	0.87	137
1945.5	75	126	0.87	119
1935.5	150	124	0.97	112
1925.5	200	124	0.94	118
1915.5	330	131	0.89	125
1905.5	130	129	0.94	124
1882.5	90	132	1.03	115
1842.5	130	132	0.92	102
1777.5	170	88*	1.09	83
1715.0	400	47*	1.03	54

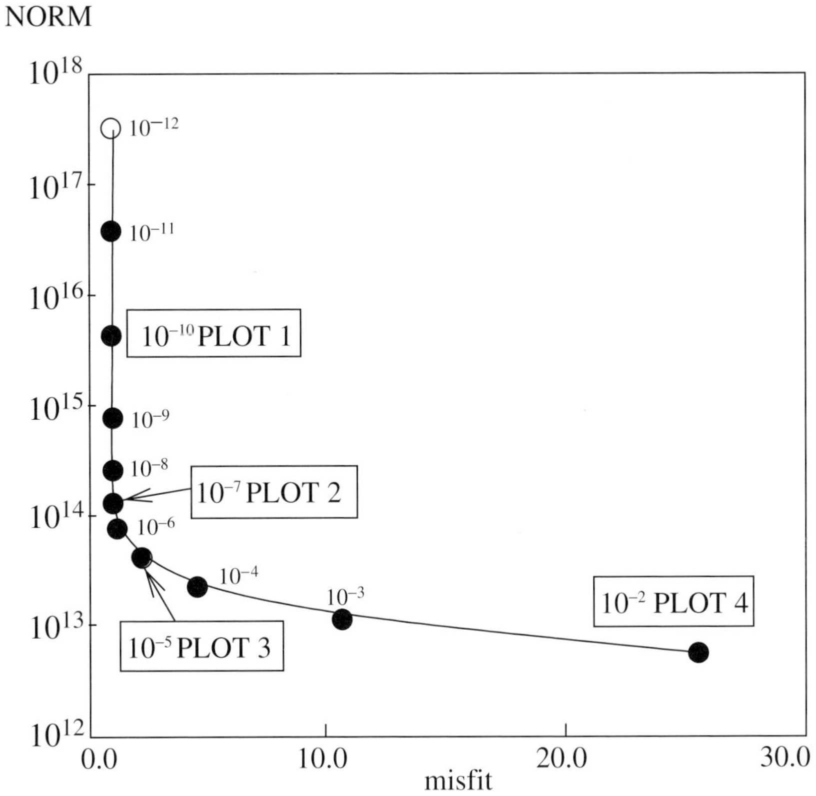

Fig. 12.3. Trade-off curve for a geomagnetic field model for epoch 1980 based on Magsat data.

log-linear and the norm is in arbitrary units. The solutions corresponding to four points on the trade-off curve are shown in Figure 12.4.

The knee lies near Point 3 with a misfit of nearly 2. The misfit never reaches unity, even for very weak damping. This suggests we have underestimated the errors or, more likely, the errors are not normally distributed. Jackson *et al.* (2000) have shown these data to have a double-exponential, rather than a Gaussian, distribution (Figure 6.6). A 1-norm solution might therefore alleviate this problem.

The solutions shown in Figure 12.4 show the effect of increased damping. Point 1 is very complicated and is dominated by small scale features, although the overall dipole structure is still clear from the preponderance of flux of different signs in the two hemispheres. Points 2 and 3 are near the knee of the trade-off curve. They have similar features but those in Point 2 are deeper. Point 4 is much smoother and contains far fewer features. These differences only show up when the models are

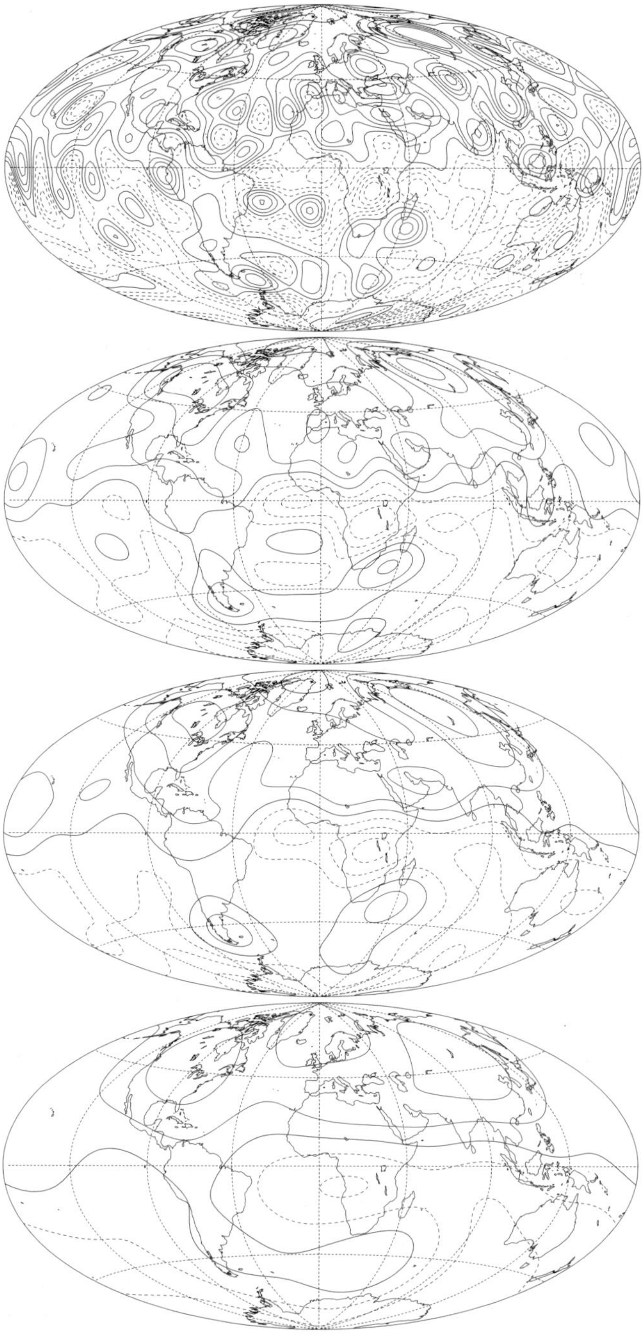

Fig. 12.4. Solutions for epoch 1980 for 4 values of the damping constant marked on the trade-off curve in Figure 12.3. B_r is plotted at the core radius. They show the effect of smoothing as the damping is increased from top to bottom. Contour interval is 100 μT in all figures.

plotted at or near the core surface: at the Earth's surface it would be hard to tell the difference.

We must now ask two rather different questions: which of these features are resolved by the data, and are they representative of the geomagnetic field at the core? The first question is the easier to answer. We have decade-long data bins and the geomagnetic field does not change much in a decade. Some features in the field have persisted for centuries. The small blue patch near Easter Island in the southeast Pacific in Point 2 appears in most epochs, even in the eighteenth century. Persistence of model features throughout a range of datasets is good evidence that the feature represents something real. The second question is much more difficult to answer. The geomagnetic field could look more like Point 1; the only control we have is the choice of damping constant, which is ultimately based on our prior belief about the geomagnetic field. If we were to construct a geodynamo model with this surface field would it have reasonable ohmic heating? Is the magnetic field in Point 4 too simple to be generated by dynamo action? These are the sort of physical questions we should now be asking.

12.7 Covariance, resolution and averaging functions

The covariance matrix applies to the geomagnetic coefficients. A point estimate on the Earth's surface (or on the core surface) is a linear combination of the geomagnetic coefficients, so its variance may be reconstructed using the quadratic form (6.53). Suppose, for example, we wish to estimate the variance of a model estimate of the vertical component at a point (θ_0, ϕ_0) on the Earth's surface. Equation (12.4) gives the estimate in terms of the geomagnetic coefficients. This has the form $\boldsymbol{a}^T\boldsymbol{x}$ that appears in (6.52), where elements of the array $\boldsymbol{a}$ are

$$(l+1)\,P_l^m(\cos\theta_0)(\cos, \sin)m\phi_0$$

The variance is then $\boldsymbol{a}^T\mathbf{C}\boldsymbol{a}$. Variances computed at the Earth's surface reflect the proximity and accuracy of the original measurements because the model is essentially an intelligent interpolation of the original data. It is 'intelligent' because it builds in the physical constraints of a potential field originating at the core surface. Variance estimates at the core surface are obtained in the same way, but array $\boldsymbol{a}$ must include the downward continuation factor $(a/c)^{l+2}$.

Figure 12.5 gives contours of standard error (square root of variance) at the core surface for three different epochs calculated in this way. In 1715.0 the large errors near the poles, particularly in the southern hemisphere, arise from paucity of data. Epoch 1915.5 had particularly good data coverage thanks to land surveys and the research vessel *Carnegie*. The 1945.5 map shows the effect of the second world war on marine surveying, with huge errors arising from lack of data in the Pacific.

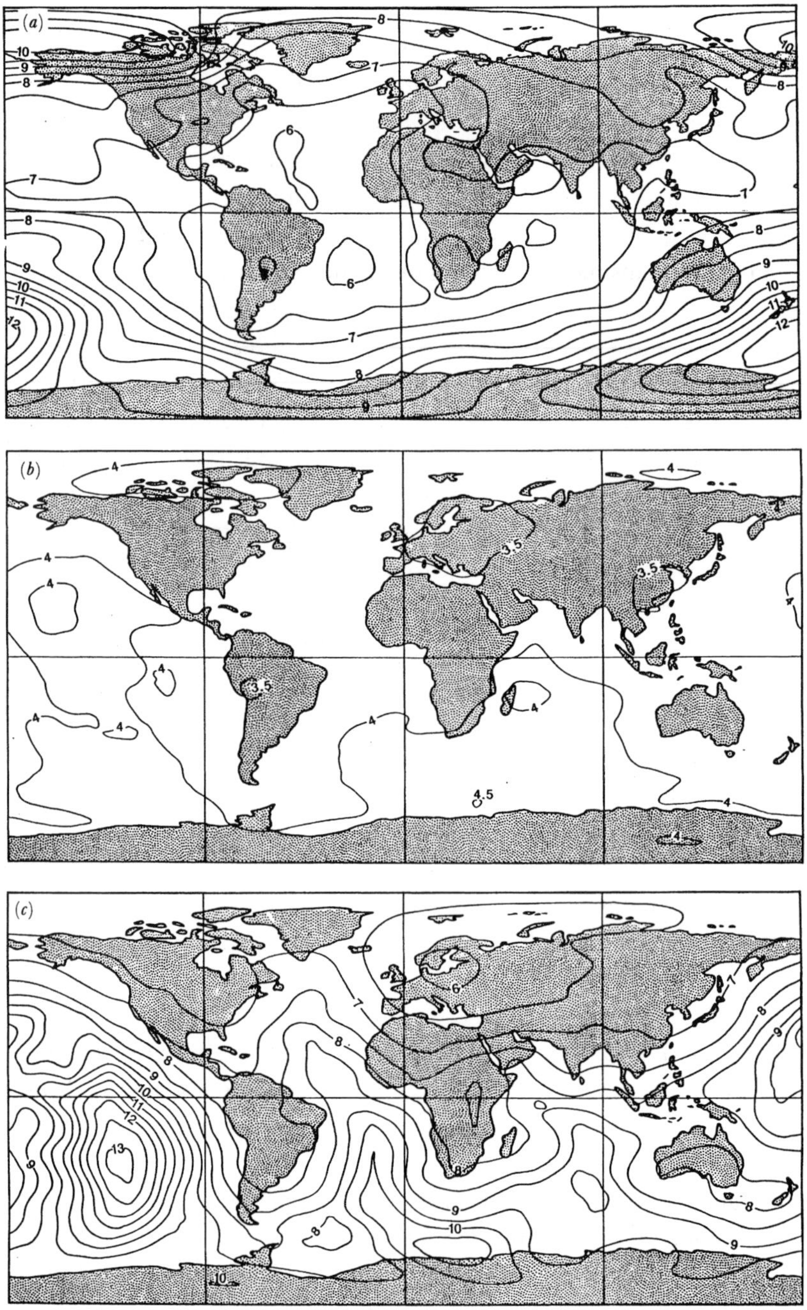

Fig. 12.5. Contour maps of the errors in the estimated radial field at the core–mantle boundary for (a) 1715.0; (b) 1915.5; and (c) 1945.5. The contour interval is 5 μT and the units are 10 μT; the projection is Plate Carré. Reproduced, with permission, from Bloxham *et al.* (1989).

The resolution matrices also apply to the geomagnetic coefficients and tell an interesting story. They are plotted for three different epochs in Figure 12.6. For perfect resolution these matrices would be diagonal. For the most recent epoch, 1966.0, this part of the matrix is quite close to diagonal and the resolution is good harmonic degree 8 or more. Only the top left-hand corner, to degree 10, has been plotted. Note that the resolution matrix is not symmetrical. Resolution is also good to degree 8 for epoch 1842.5, surprisingly so for such an old dataset, although resolution deteriorates more rapidly in the degree 9 and 10 coefficients than for the 1966.0 dataset. The oldest model, epoch 1715.0, begins to deteriorate at degree 5 and coefficients above degree 8 are virtually unresolved.

Table 12.3 contains a column for resolution. This is the trace of the resolution matrix, the sum of its diagonal elements. For perfect resolution this would equal the number of coefficients, or $L(L+2)$ where L is the maximum spherical harmonic degree. The trace of R is a measure of the number of degrees of freedom in the model, or roughly the number of resolved parameters. For epoch 1715.0 the resolution is 47, corresponding to resolving parameters to spherical harmonic degree 5 or 6. Resolution for the remaining models is remarkably uniform, showing the excellent geographical coverage of these datasets (but it says little about their accuracy). Resolution for epoch 1842.5 is 132, corresponding to harmonics up to degree 10 or 11, and for 1966.0 it is 144, about degree 11. These numbers are in agreement with plots of the resolution matrices in Figure 12.6, although the plots give a much clearer idea of where resolution fails than a single number can.

The ramping evident in the diagonal elements of the last two plots shows that coefficients with low m are less well resolved than those with high m at the same l. This is because the datasets in these early times are dominated by declination, which samples high m coefficients. Equation (12.9) shows why this is so. For a near-axial dipole field such as that of the Earth we have $Y \approx 0$ and δD nearly proportional to δY, the change in the easterly component or the ϕ-derivative of the potential. Equation (12.3) shows that differentiating with respect to ϕ brings down an m, making the partial derivative largest for large m.

The diagonal elements in the 1966.0 resolution matrix ramp the other way for degrees 9 and 10: they are largest for $m = 0$ and decrease with increasing m. This dataset is dominated by satellite measurements of total (scalar) intensity F. The ramping reflects the 'Backus effect' of the nonuniqueness left by F-data alone; the null space is described by spherical harmonic series dominated by coefficients with $l = m$.

One row of the resolution matrix gives the averaging function. The estimate of any one coefficient is a linear combination of all the coefficients, and the averaging function gives the weights applied to each. It is instructive to transform this

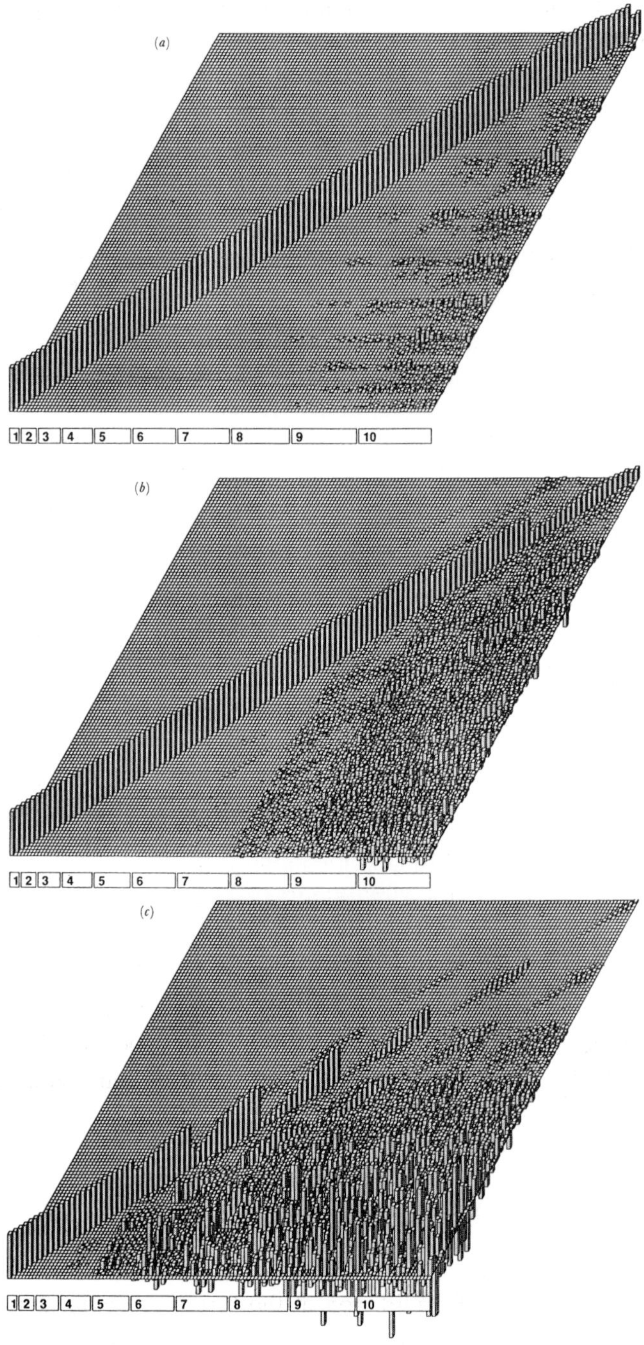

Fig. 12.6. Resolution matrices for three epochs with quite different datasets. (a) Epoch 1966.0, (b) 1842.5, and (c) 1715.0. Reproduced, with permission, from Bloxham *et al*. (1989).

averaging function into the space domain. The estimate of B_r at point (θ_0, ϕ_0) is related to the true model by

$$\hat{B}_r(\theta_0, \phi_0) = \oint A(\theta_0, \phi_0; \theta, \phi) B_r(\theta, \phi) \mathrm{d}S, \tag{12.18}$$

where A is the averaging function and the integral is taken over a spherical surface. Note that A depends on two positions on the sphere. The radial field is a linear combination of the model parameters and the estimated model is related to the true model via the resolution matrix as in (7.29). It follows that

$$\hat{B}_r(\theta_0, \phi_0) = \boldsymbol{a}^{\mathrm{T}} \boldsymbol{m} = \boldsymbol{a}^{\mathrm{T}} \mathsf{R} \boldsymbol{m}_t. \tag{12.19}$$

Orthogonality of the spherical harmonics gives $\boldsymbol{m}_t$ as an integral of $B_r(\theta_0, \phi_0)$. For example, the component g_l^m of $\boldsymbol{m}_t$ is

$$g_l^m = -\frac{l+1}{4\pi(2l+1)} \oint B_r(\theta, \phi) P_l^m(\cos\theta) \cos m\phi \mathrm{d}S = \oint c_l^m(\theta, \phi) B_r(\theta, \phi) \mathrm{d}S. \tag{12.20}$$

Define vector $\boldsymbol{c}$ to have components c_l^m. Substituting (12.20) for each of the components of $\boldsymbol{m}_t$ in (12.19) leaves an integral of the form (12.18); comparing the two gives the averaging function as simply

$$A(\theta_0, \phi_0; \theta, \phi) = \boldsymbol{a}^{\mathrm{T}}(\theta_0, \phi_0) \mathsf{R} \boldsymbol{c}(\theta, \phi). \tag{12.21}$$

If we wish the averaging function to apply between fields on the core surface, rather than the Earth's surface, the arrays $\boldsymbol{a}$ and $\boldsymbol{c}$ must contain the downward continuation factors $(a/c)^l$. Further details are in Bloxham *et al.* (1989).

Averaging functions are plotted in Figure 12.7 for three models for estimates at two different locations on the core surface. The point beneath Europe (50° N, 0° E) is well sampled at all times. There is only a slight broadening of the central peak in early times. The other point has poor data coverage in early times. This is reflected by the very diffuse averaging function in epoch 1715.0, although the other early model has excellent resolution because of the surveying activity in the Southern Ocean and Antarctica at this time.

12.8 Finding fluid motion in the core

Slow changes in the geomagnetic field are called the *secular variation*. They are caused by the motion of liquid iron at the top of the core, and secular variation may be used to estimate some properties of the fluid flow. This is an inverse problem. Progress is made only by neglecting the effects of electrical resistance, the argument being that the observed geomagnetic changes are rapid compared with the time taken for resistive effects to change the magnetic field. Alfven's

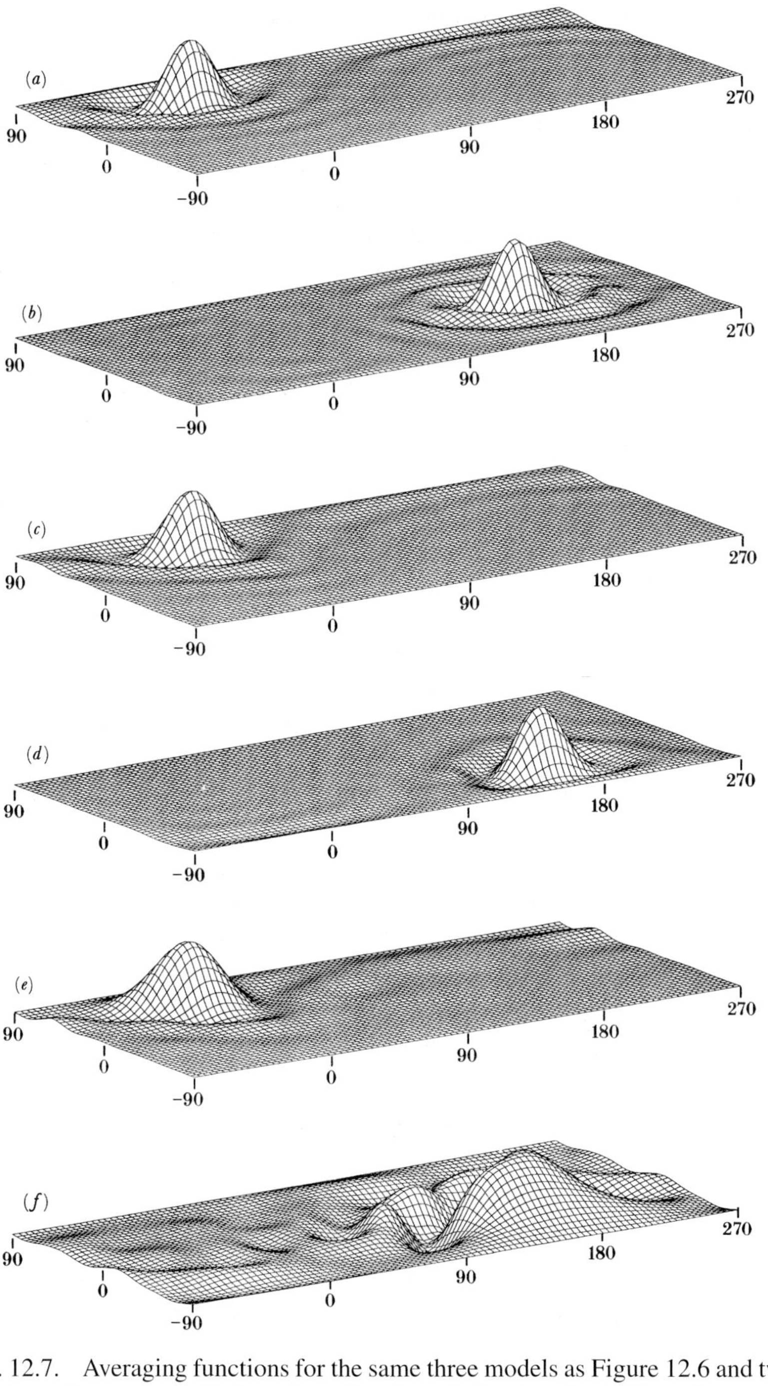

Fig. 12.7. Averaging functions for the same three models as Figure 12.6 and two points on the core–mantle boundary. Good resolution is reflected in a sharp peak centred on the chosen site. Resolution is very poor in the south Pacific for epoch 1715.0. Reproduced, with permission, from Bloxham *et al.* (1989).

theorem states that magnetic field lines are frozen into a perfect electrical conductor: they never move through the conductor but move with it. They could therefore be used to trace the flow if only we could label each field line. Unique labelling is impossible and as a result the inverse problem has a large null space. For example, consider a uniform magnetic field and a uniform translation of the fluid. There is no change in magnetic field because it remains uniform to the observer; the uniform translation therefore forms part of the null space.

We deal only with the radial component of magnetic field at the core surface because horizontal components can be affected by boundary layers and may therefore not be representative of the magnetic field in the main body of fluid near the top of the core. Contours of zero B_r play a special role in this inverse problem because they separate regions of inward-directed flux from regions of outward-directed flux. These curves must be material lines, i.e. they remain fixed to the fluid and can be used as tracers. This means we can determine the flow normal to the contours where $B_r = 0$ but not the component parallel to it, as this component does not move the line. Another part of the null space is therefore flow parallel to curves of $B_r = 0$. The total integrated flux through one of these curves must remain constant because field lines cannot cross the bounding curve. These integrals define some average properties of the flow and provide consistency conditions the data must satisfy for resistance to be neglected. Backus (1968) made an extensive study of the inverse problem and derived the conditions necessary for existence of solutions, defined the null space, and showed that the necessary conditions are sufficient to determine solutions for the core flow.

Backus' results of nonuniqueness discouraged further work on the problem for 20 years. New efforts began following the great improvement in data coverage provided by the Magsat satellite in 1980, with further constraints being placed on the flow. These included arbitrary damping, steady flow, geostrophic flow (i.e. dynamics dominated by rotation), and stratified flow (no radial motion anywhere). Here we discuss only stratified flow because it provides a rather simple illustration of this particular inverse problem.

The radial component of flow is always zero at the core surface because the liquid cannot penetrate the solid mantle. Upwelling from deeper down manifests itself by an apparent source of fluid at the surface, with flow directed radially away from the centre of upwelling. If the fluid is incompressible and there is no flow from deeper down, the horizontal divergence of the flow on the surface is zero. The divergence of an incompressible flow is zero. In cartesian coordinates with z downward, the divergence is

$$\nabla \cdot \boldsymbol{v} = \frac{\partial v_x}{\partial x} + \frac{\partial v_y}{\partial y} + \frac{\partial v_z}{\partial z} = 0. \tag{12.22}$$

At the surface we have $v_z = 0$ and if there is no vertical flow further down we have also $\partial v_z/\partial z = 0$. Equation (12.22) then shows that the horizontal divergence will be zero for stratified flow

$$\nabla_h \cdot \boldsymbol{v} = \frac{\partial v_x}{\partial x} + \frac{\partial v_y}{\partial y} = -\frac{\partial v_z}{\partial z} = 0. \tag{12.23}$$

Similar results apply in spherical coordinates with r replacing z; the equations are a little more complicated.

The induction equation gives the rate of change of the magnetic field with time in terms of inductive and resistive effects. For toroidal flow and no resistive effects the radial component of the induction equation reduces to

$$\frac{\partial B_r}{\partial t} = \boldsymbol{v} \cdot \nabla_h B_r. \tag{12.24}$$

The secular variation on the left-hand side forms the data for our inverse problem and the velocity $\boldsymbol{v}$ on the right-hand side forms the model. The partial derivatives depend on gradients of B_r. It is clear from this equation that the null space consists of flows perpendicular to $\nabla_h B_r$, or parallel to contours of B_r. Denoting the flow normal to contours of B_r by v_n, the solution of (12.24) is simply

$$v_n = \frac{1}{|\nabla_h B_r|} \frac{\partial B_r}{\partial t}. \tag{12.25}$$

There is a problem near the stationary points of B_r where $\nabla_h B_r = 0$; this provides a consistency condition on the data that $\partial B_r/\partial t = 0$ at these points.

The inverse problem can be reformulated by integrating (12.24) over the surfaces bounded by contours of B_r. The divergence theorem applies in two dimensions on the surface of a sphere, so

$$\int_S \nabla_h \cdot \boldsymbol{v} \, dS = \oint v_n \, dl, \tag{12.26}$$

where S is any patch of the core surface, the line integral is taken around its boundary, and subscript 'n' denotes the component along its normal. Integrating (12.24) over a patch bounded by a curve $B_r =$ constant gives

$$\int_S \frac{\partial B_r}{\partial t} dS = B_r \oint v_n \, dl. \tag{12.27}$$

We could, therefore, use the integrals on the left-hand side as our data. This might be preferable as it is easier to make estimates of integrals of secular variation over patches rather than at single points.

Interestingly, the integral form (12.27) yields an inverse problem that is very similar to seismic tomography because the right-hand side contains line integrals of the model. In this case the 'rays', the contours over which the line integrals are taken, are contours of B_r; they are therefore closed and never intersect. The large null space of the core motion problem could therefore be viewed as arising from the lack of any intersecting 'ray paths'. Finally, note from (12.27) that when $B_r = 0$ there is a consistency condition on the data: integrals of the secular variation around contours of zero radial magnetic field must vanish. Toroidal flows were first investigated by Whaler (1980).

In practice, the inverse problem for core motions has nearly always been solved by discretising all variables on the core surface using spherical harmonics to leave a standard linear discrete inverse problem. Stratified flow may be represented by a single scalar streamfunction ψ: $\boldsymbol{v} = \nabla \times \psi \hat{\boldsymbol{r}}$, where $\hat{\boldsymbol{r}}$ is a unit radial vector. The spherical harmonic series for B_r and its time derivative have themselves been derived from inverse problems and will have model covariances, but these have been seldom used for core motion problems. Secular variation is estimated by differencing field models from different epochs, or inverting permanent magnetic observatory data directly, and will have much larger relative errors than B_r or even the gradients of B_r that enter the equations of condition matrix. Errors in ∇B_r are ignored in core-motion inversions.

Truncation of the spherical harmonic series presents an interesting problem. Suppose we truncate $\boldsymbol{v}$ at degree N_v and B_r at N_b. The product on the right-hand side of (12.24) produces harmonics up to degree $N_v + N_b$, which means we should truncate the secular variation series at this level. Secular variation is poorly known, and we cannot hope to determine the spherical harmonic coefficients to degree N_b with any accuracy, let alone higher coefficients. There is also the inherent inconsistency of having high-degree coefficients change with time yet not be part of the magnetic field model. The correct solution is to keep the truncation of $\boldsymbol{v}$ and B_r at the value required by the prior information of the problem and set high harmonics of secular variation data to zero *with appropriate error estimates*. This prohibits unrealistic solutions for $\boldsymbol{v}$ that generate huge secular variation in very high harmonics, which can arise if these high secular variation harmonics are omitted from the inversion. Realistic error estimates allow solutions for $\boldsymbol{v}$ that fit the known harmonics without being overly constrained to force the higher harmonics to zero.

A standard damped least-squares solution to the inverse problem (12.24) provides a more stable and uniform estimate of the velocity than the simple division in (12.25), but damping produces some undesirable effects associated with the null space. This example illustrates rather dramatically the problems that arise when applying standard damping techniques to an inverse problem with a large null space. Solving (12.24) by least squares with little or no damping gives a solution that is completely dominated by the null space, and will result in large amplitude flows that are everywhere parallel to contours of B_r. Damping progressively removes components of the null space and leaves only that part of the model required for fitting the data. This will eventually leave us with v_{n}: severe damping produces flows normal to the contours of B_{r}, or as near to normal as is allowed by the geometry of the contours. This flow is everywhere perpendicular to the one at zero damping! Arbitrary damping is no solution to evaluating the importance of models in the null space; the correct solution is to find more data, or more prior information from the physics of the problem.

Another interesting effect of damping arises from secular variation in the Pacific region. The magnetic field has been remarkably constant within the Pacific basin throughout the historical period. Good core motion inversions have little flow in the Pacific because little flow is needed there. Excessively damped solutions 'spread' the model into the Pacific, putting flow where no secular variation exists. This runs contrary to our expectation of the effects of damping: that a minimum norm solution introduces only features into the model that are needed to explain the data. This is another undesirable effect of a large null space.

The special case of toroidal flows is given here to simplify the mathematics. The full problem is no more difficult to understand, and contains even more severe uniqueness problems because the allowed flows have the extra degree of freedom of radial motion. Inversion is used in core motion studies as a mapping tool: there are no pretensions at error estimates on the flows; its value lies in guiding our ideas on core dynamics rather than in giving a precise magnitude and direction at a particular point on the core surface.

Appendix 1

Fourier series

A1.1 Definitions

A *periodic function* of time, t, obeys the equation

$$a(t + T) = a(t) \tag{A1.1}$$

where T is the *period* and

$$\omega = \frac{2\pi}{T}$$

is the *angular frequency*. Any periodic function can be represented as a *Fourier series*,

$$a(t) = \frac{1}{2}a_0 + \sum_{n=1}^{\infty}(c_n \cos n\omega t + s_n \sin n\omega t), \tag{A1.2}$$

where the $\{c_n; s_n\}$ are the *Fourier coefficients*. Although $a(t)$ can have discontinuities, it must satisfy the condition

$$\int_0^T |a(t)| \, \mathrm{d}t < \infty. \tag{A1.3}$$

The cosine and sine functions are *orthogonal*, meaning

$$\int_0^T \cos n\omega t \cos m\omega t \mathrm{d}t = \frac{1}{2}T\delta_{nm} \tag{A1.4}$$

$$\int_0^T \sin n\omega t \sin m\omega t \mathrm{d}t = \frac{1}{2}T\delta_{nm} \tag{A1.5}$$

$$\int_0^T \sin n\omega t \cos m\omega t \mathrm{d}t = 0, \tag{A1.6}$$

where δ_{nm} is the *Kronecker delta*. They are also orthogonal to the function $\frac{1}{2}$ (prove it). Any reasonable function (i.e. one satisfying (A1.3)) can be expanded in the set of functions $\{\frac{1}{2}; \cos n\omega t; \sin n\omega t\}$ – the set is said to be *complete*.

To find the coefficients for any $a(t)$ from equations (A1.4)–(A1.6) do the following.

(i) Multiply by the corresponding function.
(ii) Integrate over the whole range $(0, T)$.
(iii) Use the orthogonality relations (A1.4)–(A1.6) to show that all integrals vanish except the one containing the coefficient corresponding to the relevant function.
(iv) Rearrange the equation to give the coefficient explicitly.

Thus

$$a_0 = \frac{2}{T}\int_0^T a(t)\mathrm{d}t, \tag{A1.7}$$

$$c_n = \frac{2}{T}\int_0^T a(t)\cos n\omega t\mathrm{d}t; \quad n = 0, 1, 2, \ldots, \tag{A1.8}$$

$$s_n = \frac{2}{T}\int_0^T a(t)\sin n\omega t\mathrm{d}t; \quad n = 1, 2, 3, \ldots. \tag{A1.9}$$

A1.2 Examples

(i) **The square wave**. This function may be defined as periodic with period T with

$$\begin{aligned} a(t) &= 1 \qquad 0 < t < T/2 \\ &= -1 \qquad T/2 < t < T. \end{aligned}$$

The Fourier series for $a(t)$ has the form of (A1.2) where the coefficients are given by (A1.7)–(A1.9). Substituting for $a(t)$ gives

$$a_0 = \frac{2}{T}\left[\int_0^{T/2}\mathrm{d}t - \int_{T/2}^{T}\mathrm{d}t\right] = \frac{2}{T}\left(\frac{T}{2} - \frac{T}{2}\right) = 0$$

and

$$\begin{aligned} c_n &= \frac{2}{T}\left\{\int_0^{T/2}\cos\frac{2n\pi}{T}t\mathrm{d}t - \int_{T/2}^{T}\cos\frac{2n\pi}{T}t\mathrm{d}t\right\} \\ &= \frac{1}{n\pi}\left\{\left[\sin\frac{2n\pi}{T}t\right]_0^{T/2} - \left[\sin\frac{2n\pi}{T}t\right]_{T/2}^{T}\right\} \\ &= 0 \end{aligned}$$

and

$$s_n = \frac{2}{T}\left\{\int_0^{T/2} \sin\frac{2n\pi}{T}t\,\mathrm{d}t - \int_{T/2}^{T} \sin\frac{2n\pi}{T}t\,\mathrm{d}t\right\}$$
$$= \frac{1}{n\pi}\left\{\left[-\cos\frac{2n\pi}{T}t\right]_0^{T/2} - \left[-\cos\frac{2n\pi}{T}t\right]_{T/2}^{T}\right\}$$
$$= \frac{1}{n\pi}(\cos n\pi + 1 + 1 - \cos n\pi)$$
$$= \frac{2}{n\pi}[1-(-1)^n].$$

The series can then be written as

$$a(t) = \frac{4}{\pi}\left[\sin\frac{2\pi t}{T} + \frac{1}{3}\sin\frac{6\pi t}{T} + \frac{1}{5}\sin\frac{10\pi t}{T} + \cdots\right].$$

The cosine and constant terms vanish: this always happens for *odd functions*, those with $a(t) = -a(-t)$; the sine terms always vanish for *even functions*, which have $a(t) = a(-t)$. The Fourier series represents a decomposition of the square wave into its component harmonics.

(ii) **A parabola**. It is left as an exercise to show, using formulae (A1.2) and (A1.7)–(A1.9), that the Fourier series of the function $t - t^2$ in the interval $0 < t < 1$ is

$$t - t^2 = \frac{1}{6} - \sum_{n=1}^{\infty} \frac{\cos 2n\pi t}{n^2\pi^2}.$$

(This requires some tedious integration by parts. It is possible to circumvent some of the integration by noting the function is symmetrical about the point $t = \frac{1}{2}$.)

(iii) **The sawtooth function**, defined by $S(t) = t$ on the interval $0 < t < 1$ and periodic repetition: $S(t+1) = S(t)$ (you need to sketch the function over several periods, say $t = 0$ to 3, to see why it is called a sawtooth). The Fourier series of $S(t)$ has the form of (A1.2) with $T = 1$. All integrals containing cosines vanish. Doing the remaining integrals yields:

$$S(t) = \frac{1}{2} - \sum_{n=1}^{\infty} \frac{\sin 2n\pi t}{2n\pi}.$$

A1.3 Complex Fourier series

The function $a(t)$ may also be expanded in complex exponentials.

$$a(t) = \sum_{n=-\infty}^{\infty} A_n \mathrm{e}^{in\omega t}. \qquad \text{(A1.10)}$$

The exponentials are orthogonal:

$$\int_0^T e^{in\omega t} e^{-im\omega t}\, dt = T\delta_{nm}. \tag{A1.11}$$

Note that orthogonality for complex functions requires one of the functions be complex conjugated (see Appendix 5). Multiplying both sides of (1.10) by $e^{-im\omega t}$ and integrating from 0 to T gives the formula for the complex coefficients A_n:

$$A_n = \frac{1}{T}\int_0^T a(t) e^{-im\omega t}\, dt. \tag{A1.12}$$

The complex Fourier series (A1.10) may be shown to be equivalent to the real form (A1.2). Using DeMoivre's theorem

$$e^{im\phi} = \cos m\phi + i \sin m\phi = (\cos\phi + i\sin\phi)^m,$$

we rearrange the real Fourier series (A1.2) into the complex form by choosing the coefficients

$$\begin{aligned} A_n &= \frac{c_n - i s_n}{2} \qquad n \geq 0 \\ &= \frac{c_n + i s_n}{2} \qquad n < 0. \end{aligned} \tag{A1.13}$$

Note that $A_{-n} = A_n^*$ if $a(t)$ is a real function.

A1.4 Some properties of Fourier series

(i) **Periodicity**. The Fourier series is periodic with the defined period; in fact the Fourier series is not completely specified until the period T is given. The Fourier series converges on the required function only in the given interval; outside the interval it will be different (see the parabola (ii) above) except in the special case when the function itself has the same period (e.g. the square wave (i) above).

(ii) **Even and odd functions**. Sines are odd functions $a(-t) = -a(t)$; the DC term a_0 and cosines are even functions $a(-t) = a(t)$. Even functions can be expanded in cosine terms plus a DC shift, odd functions can be expanded in sine terms.

(iii) **Half-period series**. A function $a(t)$ defined only on the interval $(0, T)$ can be expressed in either sine or cosine terms. To see this, consider the function extended to the interval $(-T, T)$ as an even function by $a(-t) = a(t)$. The Fourier series on the interval $(-T, T)$ with period $2T$ will contain only cosine terms but will converge on the required function in the original interval $(0, T)$. Similarly, extending as an odd function with $a(-t) = -a(t)$ will yield a Fourier series with only sine terms.

(iv) **Least-squares approximation**. When a Fourier series expansion of a continuous function $a(t)$ is truncated after N terms it leaves a least-squares approximation to the original function: the area of the squared difference between the two functions

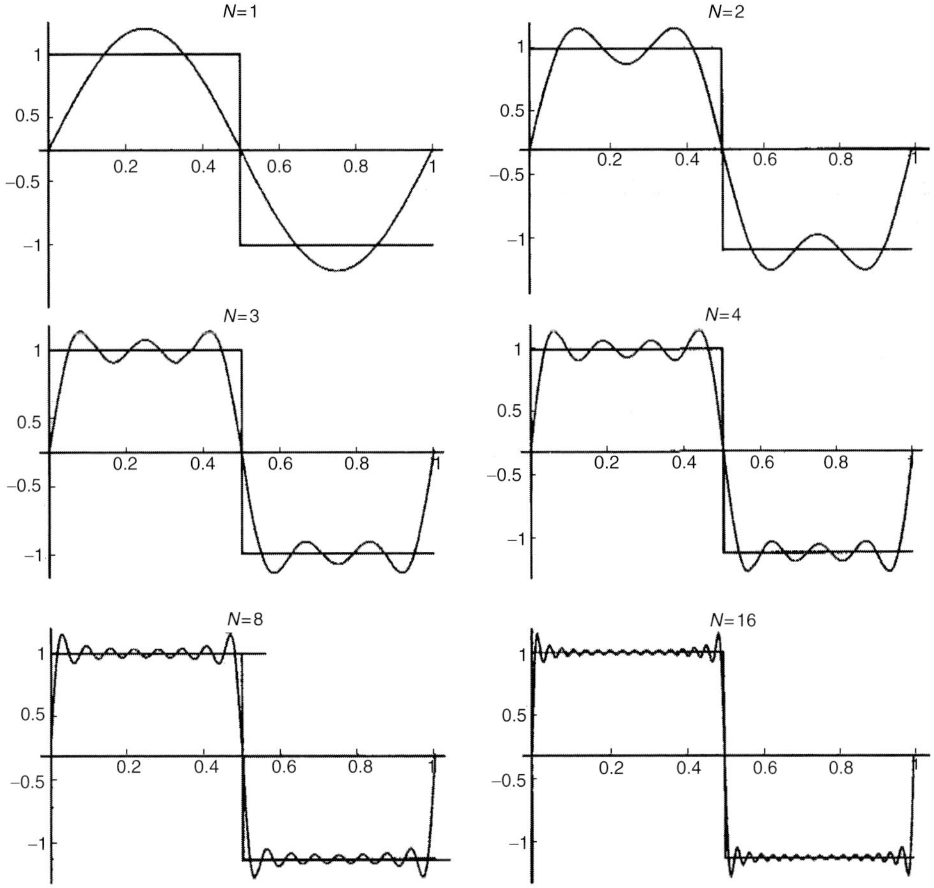

Fig. A1.1. Fourier series approximation to a square wave. The Fourier series has been truncated at $N = 1, 2, 4, 8$ and 16 terms. Note the overshoot and compression of the oscillations around the discontinuity.

is minimised and goes to zero as N increases. This property follows from orthogonality of sines and cosines: it is useful because increasing N does not change the existing coefficients; only the higher terms need be computed in order to improve the approximation. If $a(t)$ is expanded in functions that are not orthogonal, for example the monomials x^n, then raising the truncation point N requires all the coefficients to be recomputed, not just the new ones. See Appendix 5, Section A5.7(ii).

(v) **Gibbs's phenomenon**. The Fourier series converges in the least-squares sense when $a(t)$ is discontinuous, but the truncated series oscillates near the discontinuity with a constant overshoot of about 19% of the height of the discontinuity. As N increases the oscillations are compressed to a progressively smaller region about the discontinuity but the maximum error remains the same. The convergence is said to be *non-uniform* because the error depends not only on N but also on t. These oscillations are known as Gibbs's phenomenon. Figure A1.1 illustrates the effect for the square wave.

Appendix 2

The Fourier integral transform

A2.1 The transform pair

The Fourier integral transform (FIT) $F(\omega)$ of a function of time $f(t)$ is defined as

$$F(\omega) = \frac{1}{2\pi} \int_{-\infty}^{\infty} f(t) e^{-i\omega t} dt. \tag{A2.1}$$

The inverse transform is

$$f(t) = \int_{-\infty}^{\infty} F(\omega) e^{i\omega t} d\omega. \tag{A2.2}$$

Substituting (1.2) into the right-hand side of (1.1) and doing the integral serves to verify the inverse formula from the definition. The choice of factor $1/2\pi$ in (1.1) is arbitrary, as is the sign of the exponent, but of course changing them requires changing the inversion formula (1.2). The alternative factor $1/\sqrt{2\pi}$ yields a symmetric tranform pair, the same factor appearing in front of the integrals in both (1.1) and (1.2).

The integral on the right-hand side exists (i.e. is finite) provided

$$\int_{-\infty}^{\infty} |f(t)| dt < \infty. \tag{A2.3}$$

The integral transform satisfies many of the same properties as the DFT (p. 34). $F(\omega)$ is a complex function and therefore contains amplitude and phase; $F(0)$ is the integral of the entire time function (divided by 2π) and is therefore a measure of the average; the shift theorem holds: $f(t - t_0) \leftrightarrow e^{-i\omega t_0} F(\omega)$; as does Parseval's theorem and the convolution theorem; the Fourier integral transform of the derivative $f'(t)$ is $i\omega F(\omega)$; and the transform of $f(-t)$ (time reversal) is $F^*(\omega)$. There is no counterpart to aliasing because there is no maximum frequency.

A2.2 Examples

(i) **Boxcar**

$$f(t) = 1 \quad |t| < T \tag{A2.4}$$

$$= 0 \quad |t| > T. \tag{A2.5}$$

The FIT is obtained simply by evaluating the integral in (1.1) to give the *sinc function*

$$F(\omega) = \frac{\sin \omega T}{\pi \omega}, \tag{A2.6}$$

which is the continuous analogue of the sequence appearing in (1.28).

(ii) **The FIT of the Gaussian** $g(t) = \mathrm{e}^{-\alpha t^2}$ presents a slightly more challenging integral,

$$G(\omega) = \frac{1}{2\pi} \int_{-\infty}^{\infty} \mathrm{e}^{-\alpha t^2} \mathrm{e}^{-\mathrm{i}\omega t} \mathrm{d}t, \tag{A2.7}$$

which is evaluated by completing the square for the exponent:

$$\alpha \left(t + \frac{\mathrm{i}\omega}{2\alpha} \right)^2 = \alpha \left(t^2 + \frac{\mathrm{i}\omega t}{\alpha} - \frac{\omega^2}{4\alpha^2} \right). \tag{A2.8}$$

Changing variables to

$$x = \sqrt{\alpha} \left(t + \frac{\mathrm{i}\omega}{2\alpha} \right) \tag{A2.9}$$

and recalling the standard definite integral

$$\int_{-\infty}^{\infty} \mathrm{e}^{-x^2} \mathrm{d}x = \sqrt{\pi} \tag{A2.10}$$

gives

$$G(\omega) = \frac{1}{2\sqrt{\pi\alpha}} \mathrm{e}^{-\omega^2/4\alpha}. \tag{A2.11}$$

This result was used on p. 26. The FIT of a Gaussian is another Gaussian. The Gaussian function can be useful in tapering and filtering because it has no side lobes.

(iii) **The exponential function** has the power spectrum of a 1-pole Butterworth filter.

$$f(t) = \mathrm{e}^{-t} \quad t > 0 \tag{A2.12}$$

$$= 0 \quad t < 0. \tag{A2.13}$$

Integration of (1.1) gives

$$F(\omega) = \frac{1}{2\pi(1 + \mathrm{i}\omega)}. \tag{A2.14}$$

The power spectrum is then

$$|F(\omega)|^2 = \frac{1}{4\pi^2 \left(1 + \omega^2\right)}. \tag{A2.15}$$

A2.3 Properties of the Fourier integral transform

	$f(t)$	$\leftrightarrow$	$F(\omega)$
Shift theorem:	$f(t-t_0)$	$\leftrightarrow$	$\mathrm{e}^{-\mathrm{i}\omega t_0}F(\omega)$
Derivative:	$\mathrm{d}f/\mathrm{d}t$	$\leftrightarrow$	$-\mathrm{i}\omega F(\omega)$
Convolution theorem:	$\int_{-\infty}^{\infty} f(t')g(t-t')\mathrm{d}t'$	$\leftrightarrow$	$F(\omega)G(\omega)$
Symmetry:	$F(t)$ Real	$\leftrightarrow$	$F(-\omega) = F^*(\omega)$
Parseval:	$\int_{-\infty}^{\infty} \|f(t)\|^2\mathrm{d}t$	$=$	$2\pi\int_{-\infty}^{\infty} \|F(\omega)\|^2\mathrm{d}\omega$

A2.4 The delta function

Condition (1.3) is very restrictive indeed. It excludes simple functions like $f(t) = 1$ and $\sin\omega t$, for example. This is a serious drawback to use of the FIT in analysing wave motion because it cannot cope with the simplest of waves, the monochromatic sine wave.

This deficiency is rectified by using the *Dirac delta function* $\delta(t)$, which plays the part of a spike in continuous time series analysis and replaces the single non-zero entry in a time sequence. The delta function is properly defined by two properties

$$\delta(t) = 0 \quad t \neq 0 \tag{A2.16}$$

and

$$\int_{-\infty}^{\infty} \delta(t) f(t)\mathrm{d}t = f(0) \tag{A2.17}$$

for all functions $f(t)$. It follows from (1.17) with $f(t) = 1$ that

$$\int_{-\infty}^{\infty} \delta(t)\mathrm{d}t = 1, \tag{A2.18}$$

and from (1.16) that the integral is unity for any time segment that includes the origin. Intuitively, we can think of the delta function as an infinitesimally thin spike of infinite amplitude, so high that the area beneath it remains unity despite its narrowness because of (1.18). Another useful property follows from (1.17) and change of variable:

$$\int_{-\infty}^{\infty} \delta(t-t_0) f(t)\mathrm{d}t = f(t_0). \tag{A2.19}$$

This *substitution property* of the delta function makes integration trivial; the delta function is something to be welcomed rather than feared.

The next bold step is to assume that the FIT equations (1.1) and (1.2) apply to functions of time that contain the delta function. Setting $f(t) = \delta(t)$ in (1.1) gives

$$\frac{1}{2\pi}\int_{-\infty}^{\infty} \delta(t)\mathrm{e}^{-\mathrm{i}\omega t}\,\mathrm{d}t = \frac{1}{2\pi}. \tag{A2.20}$$

If Fourier's theorem applies, we can put the right-hand side of (1.20) into the inverse transform formula (1.2) and write

$$\int_{-\infty}^{\infty} \mathrm{e}^{\mathrm{i}\omega t}\,\mathrm{d}\omega = 2\pi\delta(t). \tag{A2.21}$$

Intuitively this seems plausible because the result of integrating the oscillatory function $\mathrm{e}^{i\omega t}$ for all ω gives zero, the oscillations cancelling out, except when $t = 0$, when $\mathrm{e}^{i\omega t} = 1$ and the integral diverges. A similar argument leads to

$$\int_{-\infty}^{\infty} \mathrm{e}^{-\mathrm{i}\omega t}\,\mathrm{d}t = 2\pi\delta(\omega) \tag{A2.22}$$

and by changing the sign of ω:

$$\int_{-\infty}^{\infty} \mathrm{e}^{\mathrm{i}\omega t}\,\mathrm{d}t = 2\pi\delta(-\omega) = 2\pi\delta(\omega). \tag{A2.23}$$

These two equations can be used to find the FIT for monochromatic waves:

$$\frac{1}{2\pi}\int_{-\infty}^{\infty} \cos\omega_0 t\,\mathrm{d}t = \frac{1}{2}[\delta(\omega-\omega_0)+\delta(\omega+\omega_0)] \tag{A2.24}$$

$$\frac{1}{2\pi}\int_{-\infty}^{\infty} \sin\omega_0 t\,\mathrm{d}t = \mathrm{i}\frac{1}{2}[\delta(\omega-\omega_0)-\delta(\omega+\omega_0)]. \tag{A2.25}$$

The delta function therefore allows us to represent line spectra mathematically; the Fourier transform of a monochromatic wave of frequency ω_0 is a pair of lines at $\pm\omega_0$, their signs being determined by the phase of the wave.

These equations allow us to take the FIT of a periodic function expanded in a Fourier series, as in Appendix 1 (A1.2), or the complex form (A1.10). Consider the simpler complex form. Taking the FIT of both sides of Appendix 1 (A1.10) and using (1.22) and (1.23) for the terms on the right-hand side gives

$$A(\omega) = \sum_{n=-\infty}^{\infty} A_n\delta\left(\frac{2\pi n}{T} - \omega\right). \tag{A2.26}$$

Any periodic function fails the integrability test (1.3) because the integral over each period is finite. The FIT does not exist by the strict definition, but it can be written in terms of delta functions. A combination of delta functions like the right-hand side of (1.26) is called a *Dirac comb*.

The delta function is a very powerful formalism for the physicist, but it is only that, a formalism that extends Fourier theory to include idealisations of common waveforms. Unlike the spike of discrete Fourier analysis, the delta function has no defined value at $t = 0$; note that the definition (1.16) and (1.17) avoids any mention of the actual value of $\delta(0)$.

The delta function is rendered harmless by integration. It obeys the usual rules of integration, most usefully integration by parts. Consider the effect of multiplying the derivative of $f(t)$ by a delta function and integrating. The substitution property (1.17) gives

$$\int_{-\infty}^{\infty} \delta(t)\frac{\mathrm{d}f}{\mathrm{d}t}\mathrm{d}t = \left.\frac{\mathrm{d}f}{\mathrm{d}t}\right|_{t=0}. \tag{A2.27}$$

Integration by parts yields

$$[\delta(t)f(t)]_{-\infty}^{\infty} - \int_{-\infty}^{\infty} \frac{\mathrm{d}\delta}{\mathrm{d}t} f(t)\mathrm{d}t. \tag{A2.28}$$

The first term vanishes from the first property of the delta function (1.16), leaving

$$\int_{-\infty}^{\infty} \frac{\mathrm{d}\delta}{\mathrm{d}t} f(t)\mathrm{d}t = -\left.\frac{\mathrm{d}f}{\mathrm{d}t}\right|_{t=0}. \tag{A2.29}$$

The *derivative of a delta function* picks out (minus) the derivative of the rest of the integrand. We can think of it intuitively as a pair of spikes: the delta function itself rises suddenly to infinity then plunges back to zero; its gradient goes suddenly from zero to $+\infty$, then to $-\infty$ before returning to zero. The effect of multiplying an ordinary function by such a double spike and integrating is to pick out the value of $f(t)$ just before $t = 0$ and subtract the value of $f(t)$ just after $t = 0$; the normalisation of the delta function (1.17) is just right to leave (minus) the derivative at $t = 0$.

A similar process leads to definitions of higher derivatives of the delta function. The delta function itself may be regarded as a derivative of the step function, because the indefinite integral gives

$$\begin{aligned}\int_{-\infty}^{t} \delta(t')\mathrm{d}t' &= 0 \qquad t < 0\\ &= 1 \qquad t > 0\\ &= H(t),\end{aligned} \tag{A2.30}$$

and differentiation is the reverse of integration

$$\delta(t) = \frac{\mathrm{d}H}{\mathrm{d}t}. \tag{A2.31}$$

Once again, we can think intuitively of the step function $H(t)$ rising abruptly from zero to one at $t = 0$ while its gradient jumps from zero to infinity and back to zero, just like the delta function.

I was taught delta functions twice, once in Physics and once in Mathematics. The mathematicians' method is better by far: delta functions are not ordinary functions but, strictly speaking, distributions. They may be integrated by using the Lebesgue definition of the integral, which we do not really need to understand. The rules are few and simple: distributions in isolation do not necessarily take specific values (so they are not functions); they yield specific values when multiplied by an ordinary function and integrated. They must not be squared or multiplied by another distribution even under the integral sign (although even this last rule is broken in Chapter 9, in the use of the Dirichlet condition). An accessible account of distributions is given by Lighthill (1996).

The physicists' treatment of delta functions was less rigorous and considerably more awkward. There are two unsatisfactory devices. The first is to define the delta function as a boxcar with width w and height $1/w$, giving it unit area, then take the limit $w \to 0$. It is then possible, with considerable effort, to perform some integrals to give results like (1.19) above. The usual Riemann definition of the integral involves dividing the range of x into small elements δx, summing the areas of the rectangles beneath the integrand, and letting $\delta x \to 0$. The result gives the correct value of the integral if the limit exists. With the boxcar definition of the delta function we have two limits, $w \to 0$ and $\delta x \to 0$, and we must take the second limit first to obtain an answer. This is tantamount to integrating the function before we have defined it!

The second fudge is to complete integrals involving functions like $e^{i\omega t}$ by assuming some damping to make $e^{i\omega t} \to 0$ as $t \to \pm\infty$. This also leaves some horrible integrals which eventually give the desired answer.

For me, the combination of mathematical rigour with utility (not having to do those complicated integrals) is a strong incentive to learn the simple rules for delta functions and use them. Papoulis (1962) gives many practical applications of the delta function, used in the proper way.

Appendix 3

Shannon's sampling theorem

To prove equation (3.4), consider the time series $a(t)$ with Fourier transform $A(\nu)$, where $\nu = \omega/2\pi$, such that

$$A(\nu) = 0 \quad |\nu| > \nu_N \tag{A3.1}$$

and the periodic function $A_P(\nu)$ with period $2\nu_N$, such that

$$A_P(\nu) = A(\nu) \quad 0 < |\nu| < \nu_N \tag{A3.2}$$

$$A_P(\nu) = 0 \qquad \nu_N < |\nu| < 2\nu_N. \tag{A3.3}$$

$A_P(\nu)$ has inverse Fourier transform $a_P(t)$. Being periodic, we can represent $A_P(\nu)$ as a complex Fourier series expansion:

$$\begin{aligned} A_P(\nu) &= \sum_{k=-\infty}^{\infty} a_{P_k} e^{\pi i k \nu/\nu_N} \\ &= \sum_{k=-\infty}^{\infty} a_{P_k} e^{2\pi i k \nu \Delta t}, \end{aligned} \tag{A3.4}$$

where we have used the definition of the Nyquist frequency: $\nu_N = 1/2\Delta t$. Inverting the FIT, noting from equation (A2.22) that

$$2\pi \int_{-\infty}^{\infty} e^{2\pi i k \nu \Delta t} e^{-2\pi i \nu t} d\nu = \delta(t - k\Delta t) \tag{A3.5}$$

gives the Dirac comb

$$a_P(t) = \frac{1}{2\pi} \sum_{k=-\infty}^{\infty} a_{P_k} \delta(t - k\Delta t). \tag{A3.6}$$

The coefficients in (A3.4) are given by the usual Fourier series integral:

$$a_{\mathrm{P}_k} = \frac{1}{2\nu_{\mathrm{N}}} \int_{-\nu_{\mathrm{N}}}^{\nu_{\mathrm{N}}} A_{\mathrm{P}}(\nu)\, \mathrm{e}^{-2\pi\mathrm{i}k\nu\Delta t} \mathrm{d}\nu. \tag{A3.7}$$

Comparing this with the inverse Fourier integral transform

$$a(t) = 2\pi \int_{-\nu_{\mathrm{N}}}^{\nu_{\mathrm{N}}} A_{\mathrm{P}}(\nu)\, \mathrm{e}^{-2\pi\mathrm{i}\nu t} \mathrm{d}\nu \tag{A3.8}$$

shows that

$$a_{\mathrm{P}_k} = \frac{\pi}{\nu_{\mathrm{N}}} a(k\Delta t) = \frac{\pi}{\nu_{\mathrm{N}}} a_k. \tag{A3.9}$$

The required Fourier transform, $A(\nu)$, is related to the periodic function $A_{\mathrm{P}}(\nu)$ through multiplication by a boxcar window from $\nu = -\nu_{\mathrm{N}}$ to $+\nu_{\mathrm{N}}$. By the convolution theorem for the FIT, this multiplication is equivalent to convolution of $a_{\mathrm{P}}(t)$ with the inverse Fourier transform of the boxcar, which is $b(t) = 2\sin(2\pi\nu_{\mathrm{N}} t)/t$:

$$a(t) = \frac{1}{\pi} \int_{-\infty}^{\infty} a_{\mathrm{P}}(t - t') \frac{\sin 2\pi\nu_{\mathrm{N}} t'}{(t')} \mathrm{d}t'. \tag{A3.10}$$

Substituting from (A3.6) and (A3.9) gives

$$a(t) = \frac{1}{2\pi\nu_{\mathrm{N}}} \sum_{k=-\infty}^{\infty} a_k \tag{A3.11}$$

$$\int_{-\infty}^{\infty} \delta(t - t' - k\Delta t) \frac{\sin 2\pi\nu_{\mathrm{N}} t'}{t'} \mathrm{d}t'$$

$$= \sum_{k=-\infty}^{\infty} a_k \frac{\sin 2\pi\nu_{\mathrm{N}}(t - k\Delta t)}{2\pi\nu_{\mathrm{N}}(t - k\Delta t)}. \tag{A3.12}$$

For a truncated sequence this sum is reduced to the required form in equation (3.4).

Appendix 4

Linear algebra

A4.1 Square matrices

(i) **The eigenvalue equation** for a square matrix A is

$$\mathsf{A}\boldsymbol{v} = \lambda \boldsymbol{v}, \tag{A4.1}$$

where λ is called the *eigenvalue* and $\boldsymbol{v}$ is called the *eigenvector*. If N is the order of the matrix, there are N eigenvalues, each with its corresponding eigenvector. The eigenvectors can be multiplied by any scalar (see equation (A4.1)) and we usually normalise them to unit length:

$$\boldsymbol{v}^{\mathrm{T}}\boldsymbol{v} = 1. \tag{A4.2}$$

(ii) **A symmetric matrix** has $\mathsf{A}^{\mathrm{T}} = \mathsf{A}$. Symmetric matrices have real eigenvalues and their eigenvectors are mutually perpendicular (or may be made so in the case of equal eigenvalues). *Positive semi-definite* symmetric matrices have real eigenvalues that are positive or zero (all positive = definite; some zero but none negative = semi-definite).

(iii) **Orthogonal matrices**. The matrix V whose columns are the eigenvectors of A is *orthogonal*,

$$\mathsf{V}^{\mathrm{T}}\mathsf{V} = \mathsf{V}\mathsf{V}^{\mathrm{T}} = \mathsf{I}, \tag{A4.3}$$

because the eigenvectors making up the columns of V are orthogonal. V represents a *rotation* or *reflection* of the coordinate axes, depending on whether $\det \mathsf{V} = \pm 1$ (swapping two columns of the matrix will change the sign of the determinant, which therefore only determines the ordering of the eigenvalues). We ignore reflections from now on.

(iv) **Diagonalisation**. The transformation

$$\mathsf{V}^{\mathrm{T}}\mathsf{A}\mathsf{V} = \Lambda \tag{A4.4}$$

gives a diagonal matrix Λ with eigenvalues of A along the diagonal. The transformation is a *rotation* of coordinate axes so that the (mutually orthogonal) eigenvectors become the new coordinate axes.

(v) **A quadratic form** is a solid geometrical surface whose most general form is

$$ax^2 + by^2 + cz^2 + 2hxy + 2gxz + 2fyz = K^2, \tag{A4.5}$$

which may be written in matrix form as

$$\boldsymbol{x}^{\mathrm{T}}\mathsf{A}\boldsymbol{x} = K^2 \tag{A4.6}$$

where

$$\mathsf{A} = \begin{pmatrix} a & h & g \\ h & b & f \\ g & f & c \end{pmatrix}. \tag{A4.7}$$

When rotating to diagonal form we just change axes, the surface remains the same. Its equation with respect to the new axes (x', y', z') becomes

$$\lambda_1 x'^2 + \lambda_2 y'^2 + \lambda_3 z'^2 = K^2, \tag{A4.8}$$

where the λs are the three eigenvalues of A. If all eigenvalues are positive equation (A4.5) describes an ellipsoid; if one eigenvalue is zero it describes a cylinder, and if two are zero it describes a pair of planes. Negative eigenvalues give open surfaces (e.g. a hyperboloid).

(vi) **Variational principle**. The quadratic form $\boldsymbol{v}^{\mathrm{T}}\mathsf{A}\boldsymbol{v}$, where $\boldsymbol{v}$ is a unit vector, is stationary when A is stationary and $\boldsymbol{v}$ is an eigenvector of A. The proof simply involves finding the extreme values of the quadratic form subject to the constraint $\boldsymbol{v}^{\mathrm{T}}\boldsymbol{v} = 1$ using the method of Lagrange multipliers (see Appendix 6).

Form the function $F(\boldsymbol{v})$:

$$F(\boldsymbol{v}) = \boldsymbol{v}^{\mathrm{T}}\mathsf{A}\boldsymbol{v} + \lambda(1 - \boldsymbol{v}^{\mathrm{T}}\boldsymbol{v}), \tag{A4.9}$$

where $(-\lambda)$ is the Lagrange multiplier. An arbitrary change in $\boldsymbol{v}$, $\delta\boldsymbol{v}$, produces a change δF in F, which is zero when F is stationary.

$$\delta F(\boldsymbol{v}) = \delta\boldsymbol{v}^{\mathrm{T}}\mathsf{A}\boldsymbol{v} + \boldsymbol{v}^{\mathrm{T}}\mathsf{A}\delta\boldsymbol{v} - \lambda\delta\boldsymbol{v}^{\mathrm{T}}\boldsymbol{v} - \lambda\boldsymbol{v}^{\mathrm{T}}\delta\boldsymbol{v}. \tag{A4.10}$$

Note that $\delta\boldsymbol{v}^{\mathrm{T}}\boldsymbol{v} = \boldsymbol{v}^{\mathrm{T}}\delta\boldsymbol{v}$ because they are transposes of each other and both are scalars. Similarly,

$$\delta\boldsymbol{v}^{\mathrm{T}}\mathsf{A}\boldsymbol{v} = (\boldsymbol{v}^{\mathrm{T}}\mathsf{A}^{\mathrm{T}}\delta\boldsymbol{v})^{\mathrm{T}}$$

and the first two terms are equal provided A is symmetric. Setting $\delta F = 0$ reduces to

$$\delta\boldsymbol{v}^{\mathrm{T}}(\mathsf{A}\boldsymbol{v} - \lambda\boldsymbol{v}) = 0. \tag{A4.11}$$

This must hold for any $\boldsymbol{v}$, which will be true only if

$$\mathsf{A}\boldsymbol{v} = \lambda \boldsymbol{v}, \tag{A4.12}$$

i.e. $\boldsymbol{v}$ is an eigenvector of A.

(vii) **Complex matrices**. Most of the properties of real matrices apply to matrices with complex elements, with one notable exception. Scalar products of complex vectors are formed with a complex conjugate. This ensures the length of the vector, the square root of its scalar product with itself, is a real quantity (Appendix 5). The *Hermitian conjugate* of any array is defined as the complex conjugate of the transpose and is written with a dagger †. For a complex vector $\boldsymbol{u}$ we have

$$|\boldsymbol{u}|^2 = (\boldsymbol{u}^{\mathrm{T}})^* \boldsymbol{u} = \boldsymbol{u}^{\dagger} \boldsymbol{u}. \tag{A4.13}$$

A *Hermitian matrix* is one that is equal to its Hermitian conjugate

$$\mathsf{H}^{\dagger} = (\mathsf{H}^{\mathrm{T}})^{\star} = \mathsf{H}. \tag{A4.14}$$

A real Hermitian matrix is symmetric. Hermitian matrices have many properties in common with symmetric matrices, notably real eigenvalues, orthogonal eigenvectors, and a variational principle.

(viii) **Non-symmetric matrices**. If A is not symmetric the eigenvector equation for its transpose yields a set of *left eigenvectors*:

$$\mathsf{A}^{\dagger} \boldsymbol{u} = \mu \boldsymbol{u}. \tag{A4.15}$$

Transposing both sides of this equation explains why they are called left eigenvectors:

$$\boldsymbol{u}^{\dagger} \mathsf{A} = \mu \boldsymbol{u}^{\dagger}. \tag{A4.16}$$

The eigenvalues of $\mathsf{A}^{\dagger}$ are the complex conjugates of those of A; the corresponding eigenvectors are orthogonal, or may be made so if there are equal eigenvalues. If A is a real matrix, its eigenvalues are real or occur in complex conjugate pairs (because they are the roots of a characteristic equation with real coefficients). A^{T} therefore shares the same eigenvalues as A.

The complete set of vectors $\boldsymbol{u}$ form the reciprocal vectors to $\boldsymbol{v}$. Let U be the column matrix of left eigenvectors $\boldsymbol{u}$. It is then possible to obtain a diagonal matrix

$$\Lambda = \mathsf{U}^{\dagger} \mathsf{A} \mathsf{V}, \tag{A4.17}$$

where Λ is the diagonal matrix of eigenvalues of A, which may be complex. While these formulae are correct, the product $\boldsymbol{u}^{\dagger} \boldsymbol{v}$ is not real. Hermitian conjugates give $\boldsymbol{u}^{\dagger} \boldsymbol{v}$, which is real; the $\boldsymbol{u}$ may be normalised so that $\boldsymbol{u}^{\dagger} \boldsymbol{v} = 1$ when they correspond to the same eigenvalue.

A4.2 Real rectangular matrices

(i) If the real rectangular matrix A has D rows and P columns it is said to be $D \times P$. A maps vectors from a P-dimensional space onto a D-dimensional space. A^{T} is $P \times D$ and maps vectors from a D-dimensional space onto a P-dimensional space. To multiply two matrices together, the second dimension of the first matrix must equal the first dimension of the second; there are no such restrictions on the other dimensions. Thus $\mathsf{A}^{\mathrm{T}}\mathsf{A}$ is an allowed multiplication and is a square, $P \times P$ matrix. $\mathsf{A}\mathsf{A}^{\mathrm{T}}$ is also an allowed multiplication and is a square, $D \times D$ matrix. $\mathsf{A}\mathsf{A}$ is not an allowed multiplication, nor is $\mathsf{A}^{\mathrm{T}}\mathsf{A}^{\mathrm{T}}$.

(ii) **Eigenvalues and eigenvectors of rectangular matrices.** The ordinary eigenvalue equation (A4.1) cannot apply to rectangular matrices because A now changes the dimension of the vector. However, it is still possible to write down meaningful eigenvalue equations. Consider the joint space of dimension $D + P$ formed by the union of spaces acted upon by both A and A^{T}. Then form the *partitioned matrix*:

$$\mathsf{S} = \left(\begin{array}{c|c} 0 & \mathsf{A} \\ \hline \mathsf{A}^{\mathrm{T}} & 0 \end{array} \right) \tag{A4.18}$$

Note that S is square, $(D + P) \times (D + P)$, and symmetric. Its eigenvalues are therefore real and its eigenvectors are orthogonal:

$$\mathsf{S}\boldsymbol{x} = \lambda \boldsymbol{x}. \tag{A4.19}$$

Write the eigenvector in the partitioned form $\boldsymbol{x}^{\mathrm{T}} = (\boldsymbol{v}^{\mathrm{T}}|\boldsymbol{u}^{\mathrm{T}})$, where $\boldsymbol{u}$ has dimension D and $\boldsymbol{v}$ has dimension P. The eigenvalue equation (A4.19) can then be expanded to

$$\mathsf{A}\boldsymbol{v} = \lambda \boldsymbol{u} \tag{A4.20}$$

$$\mathsf{A}^{\mathrm{T}}\boldsymbol{u} = \lambda \boldsymbol{v}. \tag{A4.21}$$

This double form is the eigenvalue equation for a rectangular matrix. Transposing (A4.21) gives

$$\boldsymbol{u}^{\mathrm{T}}\mathsf{A} = \lambda \boldsymbol{v}^{\mathrm{T}}, \tag{A4.22}$$

which shows why $\boldsymbol{u}$ is called a *left eigenvector* of A; $\boldsymbol{v}$ is called a *right eigenvector* of A.

Combining equations (A4.20) and (A4.21) gives

$$\mathsf{A}^{\mathrm{T}}\mathsf{A}\boldsymbol{v} = \lambda^2 \boldsymbol{v} \tag{A4.23}$$

$$\mathsf{A}\mathsf{A}^{\mathrm{T}}\boldsymbol{u} = \lambda^2 \boldsymbol{u}. \tag{A4.24}$$

Thus all eigenvalues of the square matrices $\mathsf{A}^{\mathrm{T}}\mathsf{A}$ and $\mathsf{A}\mathsf{A}^{\mathrm{T}}$ are positive or zero (because λ is real). $\boldsymbol{u}$ is the corresponding eigenvector of the $D \times D$ matrix $\mathsf{A}\mathsf{A}^{\mathrm{T}}$ and $\boldsymbol{v}$ is the corresponding eigenvector of the $P \times P$ matrix $\mathsf{A}^{\mathrm{T}}\mathsf{A}$. However, the two square matrices we formed from A are different sizes and therefore have different numbers of eigenvalues. In the case $D > P$ then $\mathsf{A}^{\mathrm{T}}\mathsf{A}$ is the smaller of the two square matrices and will have P positive (or possibly zero) real eigenvalues. AA^T is the larger of the two matrices; it will have the same P eigenvalues as the small matrix and an additional $D - P$ real eigenvalues with orthogonal eigenvectors. Let one of these eigenvectors be $\boldsymbol{u}_1$; then by orthogonality

$$\boldsymbol{u}_1^{\mathrm{T}} \mathsf{A} \boldsymbol{v} = 0 \tag{A4.25}$$

for all P linearly independent eigenvectors $\boldsymbol{v}$. Transposing this equation gives

$$\boldsymbol{v}^{\mathrm{T}} \mathsf{A}^{\mathrm{T}} \boldsymbol{u}_1 = 0. \tag{A4.26}$$

This equation holds for all the vectors $\boldsymbol{v}$, which span the P-dimensional space, and therefore

$$\mathsf{A}^{\mathrm{T}} \boldsymbol{u}_1 = 0 \tag{A4.27}$$

and obviously

$$\mathsf{AA}^{\mathrm{T}} \boldsymbol{u}_1 = 0. \tag{A4.28}$$

The remaining $D - P$ eigenvalues are therefore zero; the corresponding eigenvectors of S have the partitioned form $(0|\boldsymbol{v})$.

In the case $P > D$ the same remarks apply in reverse: $\mathsf{A}^{\mathrm{T}}\mathsf{A}$ is the larger of the two matrices and has $P - D$ zero eigenvalues.

A4.3 Generalised inverses

(i) **The null space**. If any vector $\boldsymbol{m}_0$ satisfies the equation

$$\mathsf{A}\boldsymbol{m}_0 = 0A, \tag{A4.29}$$

then $\boldsymbol{m}_0$ is a right eigenvector of A with zero eigenvalue; it is also an eigenvector of $\mathsf{A}^{\mathrm{T}}\mathsf{A}$, which will have a zero eigenvalue corresponding to this eigenvector even in the case $P < D$. The set of all vectors satisfying (A4.29) spans a vector space of dimension $N \leq P$ called the *null space*. The general vector belonging to this set is sometimes called the *annihilator* of A.

A square matrix with zero eigenvalue has no inverse in the usual sense, i.e. B is the inverse of A if

$$\mathsf{AB} = \mathsf{BA} = \mathsf{I}. \tag{A4.30}$$

It is also impossible to solve a system of linear equations of the type

$$\mathsf{A}\boldsymbol{m} = \boldsymbol{d} \tag{A4.31}$$

in the usual way. However, this system of equations has many solutions because it is deficient. If one solution $\boldsymbol{m}$ can be found, then any vector of the form $\boldsymbol{m} + \alpha \boldsymbol{m}_0$ will also be a solution, where α is any scalar multiplier. $\boldsymbol{m}_0$ defines an uncertainty in the solution. The null space defines the full uncertainty in the solution; the complete solution to (A4.31) is a single vector orthogonal to the null space plus an arbitrary linear combination of all vectors in the null space.

Even when there are no null vectors $\boldsymbol{m}_0$, and $D > P$, the matrix AA^{T} has a null space of dimension $N = D - P$ because of the zero eigenvalues found in (A4.28).

(ii) **Least-squares inversion of rectangular matrices.** Rectangular and singular square matrices do not have inverses in the normal sense, and it is useful to generalise the concept of inverse of a matrix by relaxing the two conditions (A4.30). The least-squares solution developed in Section 6.2 is one example of a generalised inverse. It satisfies only one of the equations (A4.30). Ignoring the error vector in (6.4) gives (A4.31); we write the solution (6.26) as

$$\boldsymbol{m} = \mathsf{A}^{-N}\boldsymbol{d}, \qquad \text{(A4.32)}$$

where

$$\mathsf{A}^{-N} = (\mathsf{A}^{\mathrm{T}}\mathsf{A})^{-1}\mathsf{A}^{\mathrm{T}}. \qquad \text{(A4.33)}$$

It is easily verified that $\mathsf{A}^{-N}\mathsf{A} = \mathsf{I}$ but AA^{-N} has no meaning because the dimensions are wrong for the multiplication.

(iii) **Matrix inversion by eigenvector expansion.** A generalised inverse can be constructed for a singular, symmetric, square matrix A in terms of its eigenvectors and eigenvalues. First rotate axes to diagonalise the matrix; equation (A4.4) yields

$$\mathsf{A} = \mathsf{V}\Lambda\mathsf{V}^{\mathrm{T}}. \qquad \text{(A4.34)}$$

The inverse of Λ is obviously the diagonal matrix with elements $1/\lambda^{(i)}$. Provided A is not singular, its inverse can be reconstructed by rotating back:

$$(\mathsf{A}^{\mathrm{T}}\mathsf{A})^{-1} = \mathsf{V}\Lambda^{-1}\mathsf{V}^{\mathrm{T}}. \qquad \text{(A4.35)}$$

The procedure does not work when an eigenvalue is zero because some elements of Λ^{-1} cannot be found. A generalised inverse can be constructed by replacing the offending elements $1/\lambda^{(i)}$ with appropriate alternatives: the usual choice is to make them zero. Writing this inverse as Λ^{-G}, the generalised inverse of A becomes

$$(\mathsf{A})^{-G} = \mathsf{V}\Lambda^{-G}\mathsf{V}^{\mathrm{T}}, \qquad \text{(A4.36)}$$

where Λ^{-G} is diagonal with elements $1/\lambda^{(i)}$ when $\lambda^{(i)} \neq 0$ and zero when $\lambda^{(i)} = 0$.

Multiplying by the original matrix then using (A4.4) and the orthogonality of V from (A4.3) gives

$$\begin{aligned}\mathsf{A}^{-G}\mathsf{A} &= \mathsf{V}^{\mathrm{T}}\Lambda^{-G}\mathsf{V}\mathsf{V}^{\mathrm{T}}\Lambda\mathsf{V} \\ &= \mathsf{V}^{\mathrm{T}}\Lambda^{-G}\Lambda\mathsf{V} = \mathsf{I}^{G},\end{aligned} \qquad \text{(A4.37)}$$

where I^{G} is a diagonal matrix with ones on the leading diagonal when Λ has a non-zero eigenvalue, zero when it has a zero eigenvalue. Any vector can be expanded in a complete set of eigenvectors of A. Multiplying by I^{G} has the effect of removing any contributions from eigenvectors in the null space of A; it is therefore a *projection operator* that projects the vector out of the null space.

(iv) **Damped inverse**. The damped inverse is similar to the generalised inverse. Instead of zeroing contributions from the null space, all the eigenvalues are increased by a fixed amount. There are no longer any zero eigenvalues and the square matrix may be inverted in the usual way. If the matrix A has eigenvalues λ_i, the matrix $(\mathsf{A} + \epsilon \mathsf{I})$ will have eigenvalues $\lambda_i + \epsilon$. Λ^{-1} in (A4.35) is replaced by Λ^{-D}, which has diagonal elements $1/(\lambda_i + \epsilon)$. When $\lambda_i = 0$ the corresponding eigenvector is multiplied by ϵ^{-1}. The damped inverse is useful when the projection of $\boldsymbol{d}$ onto the null space is not too large, so that ϵ can be chosen to be smaller than most typical eigenvalues while at the same time the amplification factor ϵ^{-1} does not increase the components in the null space too much.

(v) **Minimum-norm solution.** Another way to solve a deficient set of equations is to select the shortest solution for $\boldsymbol{m}$, the one with the minimum length or norm. The generalised inverse A^{-G} already does this by projecting the solution vector out of the null space; there is no unnecessary component and the solution must be the shortest possible.

The following alternative derivation of the minimum-norm solution is instructive. Minimise the length squared, $\boldsymbol{m}^{\mathrm{T}}\boldsymbol{m}$, subject to the constraints imposed by the given equations $\mathsf{A}\boldsymbol{m} = \boldsymbol{d}$ using the method of Lagrange multipliers. Arrange the Lagrange multipliers into a vector $\boldsymbol{\lambda}$ and minimise the quantity

$$\mathcal{L} = \boldsymbol{m}^{\mathrm{T}}\boldsymbol{m} + \boldsymbol{\lambda}^{\mathrm{T}}(\mathsf{A}\boldsymbol{m} - \boldsymbol{d}) \tag{A4.38}$$

by differentiating with respect to $\boldsymbol{m}$, which gives

$$\frac{\partial \mathcal{L}}{\partial m_j} = 2m_j + \sum_{i=1}^{D} \lambda_i A_{ij} = 0. \tag{A4.39}$$

Rearranging gives

$$\boldsymbol{m} = -\frac{1}{2}\mathsf{A}^{\mathrm{T}}\boldsymbol{\lambda}. \tag{A4.40}$$

The Lagrange multipliers are found from the constraint

$$\boldsymbol{d} = \mathsf{A}\boldsymbol{m} = -\frac{1}{2}(\mathsf{A}\mathsf{A}^{\mathrm{T}})\boldsymbol{\lambda}. \tag{A4.41}$$

This time, if $P > D$ the square matrix $\mathsf{A}\mathsf{A}^{\mathrm{T}}$ has the smaller dimension and will (usually) have all D eigenvalues non-zero. Its inverse exists, and (A4.41) can be solved in the form

$$\boldsymbol{\lambda} = -2(\mathsf{A}\mathsf{A}^{\mathrm{T}})^{-1}\boldsymbol{d}. \tag{A4.42}$$

Substituting from (A4.40) gives the minimum length solution for the model

$$\boldsymbol{m} = \mathsf{A}^{\mathrm{T}}(\mathsf{A}\mathsf{A}^{\mathrm{T}})^{-1}\boldsymbol{d}. \tag{A4.43}$$

Just as the least-squares solution gave the generalised inverse for the rectangular matrix A when $P < D$, the minimum-norm solution gives a generalised inverse for the case $P > D$:

$$\mathsf{A}^{-M} = \mathsf{A}^{\mathrm{T}}(\mathsf{A}\mathsf{A}^{\mathrm{T}})^{-1}. \tag{A4.44}$$

Note that $\mathsf{A}\mathsf{A}^{-M} = \mathsf{I}$, whereas $\mathsf{A}^{-M}\mathsf{A}$ is not defined; the opposite was the case for A^{-N} defined in (A4.33).

Appendix 5

Vector spaces and the function space

The elementary concept of a vector is usefully extended to N dimensions, where N can be greater than three or infinite.

A5.1 Linear vector space

A *linear vector space* $\mathcal{V}$ is a set of elements $\boldsymbol{a}, \boldsymbol{b}, \boldsymbol{c}$ which combine $(+)$ under the rules

- closure: $\boldsymbol{a} + \boldsymbol{b} \subset \mathcal{V}$ is a vector in $\mathcal{V}$
- commutation: $\boldsymbol{a} + \boldsymbol{b} = \boldsymbol{b} + \boldsymbol{a}$
- association: $(\boldsymbol{a} + \boldsymbol{b}) + \boldsymbol{c} = \boldsymbol{a} + (\boldsymbol{b} + \boldsymbol{c})$
- null vector exists and is unique: $\boldsymbol{a} + \boldsymbol{0} = \boldsymbol{a}$
- the negative $-\boldsymbol{a}$ exists: $\boldsymbol{a} + (-\boldsymbol{a}) = \boldsymbol{0}$

and can be multiplied by scalars (α, β) under the rules

- closure: $\alpha\boldsymbol{a} \subset \mathcal{V}$
- distribution: $\alpha(\boldsymbol{a} + \boldsymbol{b}) = \alpha\boldsymbol{a} + \alpha\boldsymbol{b}$
- association: $(\alpha + \beta)\boldsymbol{a} = \alpha\boldsymbol{a} + \beta\boldsymbol{a}$.

A5.2 Dimension

A set of vectors $\{\boldsymbol{a}, \boldsymbol{b}, \boldsymbol{c}, \ldots\}$ in $\mathcal{V}$ is *linearly independent* if the equation

$$\alpha\boldsymbol{a} + \beta\boldsymbol{b} + \gamma\boldsymbol{c} + \cdots = 0 \tag{A5.1}$$

implies $\alpha = \beta = \gamma = \cdots = 0$. If there exist N linearly independent vectors but no such set of $N + 1$ vectors then N is the *dimension* of the space.

A5.3 Basis vectors and coordinates

A set of N linearly independent vectors $\{\boldsymbol{e}^{(1)}, \boldsymbol{e}^{(2)}, \cdots + \boldsymbol{e}^{(N)}\}$ is said to *span* and form a *basis* for the space of dimension N. Any vector $\boldsymbol{x} \subset \mathcal{V}$ may be written as a linear combination of the basis vectors:

$$\boldsymbol{x} = x_1\boldsymbol{e}^{(1)} + x_2\boldsymbol{e}^{(2)} + \cdots + x_N\boldsymbol{e}^{(N)}. \tag{A5.2}$$

The $\{x_i\}$ are the *components* of the vector $\boldsymbol{x}$ in the coordinate system formed by this basis.

A5.4 Scalar product

The scalar product of $\boldsymbol{a}$ and $\boldsymbol{b}$, $\boldsymbol{a} \cdot \boldsymbol{b}$, satisfies the following rules:

- $\boldsymbol{a} \cdot \boldsymbol{b} = \boldsymbol{b} \cdot \boldsymbol{a}$
- $\boldsymbol{a} \cdot (\alpha\boldsymbol{b} + \beta\boldsymbol{c}) = \alpha\boldsymbol{a} \cdot \boldsymbol{b} + \beta\boldsymbol{a} \cdot \boldsymbol{c}$
- $\boldsymbol{a} \cdot \boldsymbol{a} \geq 0$
- $\boldsymbol{a} \cdot \boldsymbol{a} = 0 \Rightarrow \boldsymbol{a} = 0$
- $(\boldsymbol{a} \cdot \boldsymbol{a})^{\frac{1}{2}}$ is the *length* of $\boldsymbol{a}$.

If $\boldsymbol{a} \cdot \boldsymbol{b} = 0$ then $\boldsymbol{a}, \boldsymbol{b}$ are said to be *orthogonal*. If the basis vectors are mutually orthogonal and of unit length,

$$\boldsymbol{e}^{(i)} \cdot \boldsymbol{e}^{(j)} = \delta_{ij}, \tag{A5.3}$$

they are said to be *orthonormal*. The components of $\boldsymbol{x}$ in an orthonormal basis are given by

$$x_i = \boldsymbol{x} \cdot \boldsymbol{e}^{(i)}. \tag{A5.4}$$

If the scalar product is defined as the sum of products of components in an orthonormal basis the vector space is called *N-dimensional Euclidean*:

$$\boldsymbol{a} \cdot \boldsymbol{b} = a_1b_1 + a_2b_2 + \cdots + a_Nb_N. \tag{A5.5}$$

A5.5 The function space

Consider a set of functions $f(x)$, $g(x)$, ..., defined on the interval (a, b). The usual Riemann definition of the integral is

$$\int_a^b f(x)g(x)\mathrm{d}x = \lim_{N\to\infty} \sum_{i=1}^{N} f(x_i)g(x_i)\Delta x, \tag{A5.6}$$

where $x_1 = a, x_N = b, x_{i+1} - x_i = \Delta x = (b - a)/(N - 1)$. Apart from the factor Δx, the right-hand side of (A5.6) is a Euclidean scalar product. The functions $f(x)$ obey the rules of combination and multiplication by a scalar and form a vector space provided enough functions are included to satisfy the closure conditions. $f(x_i)$ may be thought of as the ith component of $f(x)$.

The dimension of the space of functions is infinite, $N = \infty$. The scalar product of two functions f and g is usually called an *inner product* and is written

$$(f, g) = \int_a^b f(x)g(x)\mathrm{d}x. \tag{A5.7}$$

A function space in which the scalar product is defined in this way is called a *Hilbert space*. The definition of the scalar product in (A5.7) may be made without reference to the Riemann definition of the integral because it satisfies the properties required of a general scalar product; equation (A5.6) is useful in drawing the comparison with the more familiar scalar product of two vectors in three-dimensional Euclidean space but is not necessary for the development of the theory of Hilbert spaces. If

$$\int_a^b f(x)g(x)\mathrm{d}x = 0 \tag{A5.8}$$

the functions f and g are said to be *orthogonal*.

A5.6 Complex vector spaces

Complex vector spaces require a change in the definition of the scalar product to make the length of a vector real. Equation (A5.5) becomes

$$\boldsymbol{a} \cdot \boldsymbol{b} = a_1^\star b_1 + a_2^\star b_2 + \cdots + a_N^\star b_N, \tag{A5.9}$$

which leaves $|\boldsymbol{a}|$ real:

$$|\boldsymbol{a}|^2 = \boldsymbol{a} \cdot \boldsymbol{a} = \sum_{i=1}^{N} |a_i|^2. \tag{A5.10}$$

The first property of the scalar product (commutation) is changed to

$$\boldsymbol{a} \cdot \boldsymbol{b} = (\boldsymbol{b} \cdot \boldsymbol{a})^\star. \tag{A5.11}$$

The same rules apply to the function space. The scalar product becomes

$$(f, g) = \int_a^b f^\star(x)g(x)\mathrm{d}x. \tag{A5.12}$$

For example, the exponentials used in complex Fourier series are orthogonal:

$$\left(e^{2\pi i n t/T}, e^{2\pi i m t/T}\right) = \int_0^T e^{-2\pi i n t/T} e^{2\pi i m t/T} dt = \int_0^T e^{2\pi i (m-n)t/T} dt$$

$$= \left[\frac{e^{2\pi i (m-n)t/T}}{2\pi i (m-n)}\right]_0^T = \delta_{mn}. \qquad \text{(A5.13)}$$

A5.7 Basis functions

The basis for a function space is an infinite set of functions. The following examples will demonstrate their use.

(i) **Fourier series**. If $a = 0, b = 2\pi$, and all the functions are periodic with period 2π, the sines, cosines, and constant function form an orthonormal basis:

$$\boldsymbol{e}^{(0)} = \frac{1}{\sqrt{2\pi}} \qquad \text{(A5.14)}$$

$$\boldsymbol{e}^{(cm)} = \frac{1}{\sqrt{\pi}} \cos mx \qquad \text{(A5.15)}$$

$$\boldsymbol{e}^{(sm)} = \frac{1}{\sqrt{\pi}} \sin mx. \qquad \text{(A5.16)}$$

The functions are orthogonal because

$$\frac{1}{\pi}\int_0^{2\pi} \cos mx \cos nx \, dx = \frac{1}{\pi}\int_0^{2\pi} \sin mx \sin nx \, dx = \delta_{nm} \qquad \text{(A5.17)}$$

and

$$\frac{1}{\pi}\int_0^{2\pi} \cos mx \sin nx \, dx = 0. \qquad \text{(A5.18)}$$

The expansion of a member of the set, $f(x)$, in the basis vectors then becomes the Fourier series of Appendix (A1.2):

$$f(x) = c_0 \times \frac{1}{\sqrt{2\pi}} + \sum_{m=1}^{\infty} c_m \frac{1}{\sqrt{\pi}} \cos mx + s_m \frac{1}{\sqrt{\pi}} \sin mx. \qquad \text{(A5.19)}$$

The components of f are given by the scalar products with the basis functions, which have the form:

$$c_m = \int_0^{2\pi} f(x) \frac{1}{\sqrt{2\pi}} \cos mx \, dx. \qquad \text{(A5.20)}$$

This is close to the standard formula for the coefficients in a Fourier series.

Note that sines alone would form a basis for the set of odd functions $f(-x) = -f(x)$. They form a basis for all periodic functions with period π but only 'half' those with period 2π. The same applies to the cosines and the set of even functions $f(-x) = f(x)$. See the discussion of half-period Fourier series in Appendix 1.

(ii) **Polynomials**. The set of monomials $\{x^m;\ m = 0, 1, 2, \ldots\}$ on the interval $(-1, 1)$ form a vector space. A few simple trial integrations show that the monomials are not orthogonal. The only set of polynomials that form an orthogonal basis for this set are the Legendre polynomials. To prove this, start from the functions $P_0(x) = 1$ and $P_1(x) = x + \alpha$ and apply the orthogonality condition to determine α. Then define $P_2(x) = x^2 + \beta x + \gamma$ and apply the two orthogonality conditions for P_0 and P_1 to find β and γ. The functions may be normalised by multiplying by another scalar found by applying the normalisation condition. For example, P_0 is a constant and the condition $\int_{-1}^{1} P_0^2(x)\mathrm{d}x = 1$ requires the constant to be $\frac{1}{2}$.

The usual form for the Legendre polynomials is not normalised, but $[P_l(x)]^2$ integrates to $2/(2l + 1)$. They are

$$\begin{aligned} P_0(x) &= 1 \\ P_1(x) &= x \\ P_2(x) &= \frac{1}{2}(3x^2 - 1) \\ \ldots & \quad . \end{aligned}$$

Legendre polynomials play an important role in the theory of polynomials and approximation theory. If a polynomial approximation is needed, it is best to use a series of Legendre polynomials rather than monomials because of their orthogonality. Among the many advantages is the same least-squares best fit property as Fourier series (see Appendix 1). If a better approximation is required by adding more terms to an expansion in Legendre polynomials, the existing terms remain unchanged. Suppose, for example, a cubic approximation $a + bx + cx^2 + dx^3$ is required instead of a quadratic $a + bx + cx^2$ approximation. The coefficients a, b, c will be different in the two cases, whereas the first three coefficients of a Legendre expansion are the same in both cases.

The functions P_l with l even are even functions $P_{2l}(-x) = P_{2l}(x)$ because they contain only even powers of x, whereas the P_{2l+1} are odd functions because they contain only odd powers of x. Like the cosines and sines, the P_{2l} form a basis for the vector space of even functions and the P_{2l+1} a basis for the space of odd functions; either form a basis for the space of functions on the interval (0, 1).

(iii) **Splines**. Splines are polynomial functions defined over a small number of sampling points or *nodes*. Adjacent splines join so that they form a continuous function with continuous first and second derivatives. They are called splines because a bent beam has the same property – continuity of the first two derivatives everywhere, with a discontinuity in third derivative wherever a bending moment is applied. The most common spline is a cubic fitted through three points. Although they are defined locally, they form a *global* representation in general because the coefficients of each cubic depend on the coefficients of all the other cubics.

The term spline is often used loosely to mean any simple function defined over a finite range. Spline representations can then be made local, and orthogonal. An important group of functions is the B-splines, which retain the desirable continuity and differentiability of cubic splines but have also the advantages of a local representation. A most intelligible account of B-splines is in Press *et al.* (1992); a comprehensive account of splines is in Schultz (1973).

A5.8 Linear operators

A *linear operator* A acts on a vector $\boldsymbol{a}$ to produce another vector $\boldsymbol{b}$,

$$\mathsf{A}\boldsymbol{a} = \boldsymbol{b}. \tag{A5.21}$$

The operator is only linear if it satisfies the property:

$$\mathsf{A}(\alpha\boldsymbol{a} + \beta\boldsymbol{b}) = \alpha\mathsf{A}\boldsymbol{a} + \beta\mathsf{A}\boldsymbol{b}. \tag{A5.22}$$

For Euclidean vector spaces of finite dimension, linear operators can be represented by matrices of components in the chosen basis. Linear operators usually map to vectors in the same space, but not necessarily so. Rectangular matrices map to a space of different dimension, for example.

Differential and integral operators act on functions in a function space. Linear differential operators have many properties in common with matrices. For example, the second derivative may be represented by the limit of the second-order finite difference,

$$\frac{\mathrm{d}^2 f}{\mathrm{d}x^2} = \lim_{\Delta x \to 0} \frac{f(x_{i+1}) - 2f(x_i) + f(x_{i-1})}{\Delta x^2} = \mathsf{A}f, \tag{A5.23}$$

which in matrix form gives rows as follows:

$$\mathsf{A}f = \begin{pmatrix} \cdots & \cdots & \cdots & \cdots & \cdots \\ 1 & -2 & 1 & \cdots & \cdots \\ 0 & 1 & -2 & 1 & \cdots \\ 0 & 0 & 1 & -2 & 1 \\ \cdots & \cdots & \cdots & \cdots & \cdots \end{pmatrix} \begin{pmatrix} \cdots \\ f(x_{i-1}) \\ f(x_i) \\ f(x_{i+1}) \\ \cdots \end{pmatrix}. \tag{A5.24}$$

For the interval $(0, 2\pi)$ and homogeneous boundary conditions $f(0) = f(2\pi) = 0$ the *eigenfunctions* of the operator $\mathrm{d}^2/\mathrm{d}x^2$ are the solutions of the differential equation

$$\frac{\mathrm{d}^2 f}{\mathrm{d}x^2} = -m^2 f. \tag{A5.25}$$

They are the sines and cosines of the Fourier series. Equation (A5.25) is analogous to the matrix eigenvalue equation $\mathsf{A}\boldsymbol{x} = \lambda \boldsymbol{x}$, with A defined as in (A5.24).

Some differential operators, called *Hermitian* or *self-adjoint* depending on whether they are complex or real, are analogues of symmetric matrices and have orthogonal eigenfunctions, like the Fourier functions here. These eigenfunctions form useful bases for many problems, just as eigenvectors form useful bases for finite-dimensional vector spaces. Note that not all approximations of Hermitian differential operators lead to symmetric matrices. For example, the first derivative of f could be approximated either by the second-order, centred-difference form $[f(x_{i+1}) - f(x_{i-1})]/2\Delta x$ or by the forward difference $[f(x_{i+1} - f(x_i)]/\Delta x$. The first form gives a symmetric matrix, the second does not. This is one example of the many properties of the continuous problem (forward or inverse) that may be lost when approximating with a discrete formulation.

Appendix 6

Lagrange multipliers and penalty parameters

A6.1 The method of undetermined multipliers

In optimisation methods we often wish to minimise (or find the extreme values of) a function of several variables $F(x, y, \ldots)$ subject to some constraint $G(x, y \ldots) = 0$. The most common example in this book comes from a linear inverse problem, where one wishes to minimise a quadratic function of the model parameters, the norm of the model, subject to a set of linear constraints, the data expressed as a linear combination of the model parameters. (It is somewhat alarming to find all the physics of the problem relegated to a side constraint like this, but it often happens!)

If the function G is sufficiently simple we can rearrange it to express one of the independent variables in terms of all the others, substitute back into F to eliminate that one variable, and perform the minimisation by differentiating with respect to the remaining variables. It is usually not possible to do this analytically, and even when it is possible it is often inelegant or inconvenient. Lagrange's method of undetermined multipliers provides a simple and elegant solution, although at first sight it seems to make the problem longer by adding extra unknowns and equations.

We introduce the new combination H

$$H(x, y, \ldots) = F(x, y, \ldots) + \lambda G(x, y, \ldots), \tag{A6.1}$$

where λ is an unknown parameter. If the constraints are satisfied then $H \equiv F$, and the minimum of H will also give the minimum of F. H is minimised by differentiating with respect to each of the N independent parameters in turn and setting the result to zero in the usual way. This gives N equations for the $N + 1$ unknowns, the values of the independent variables, and the Lagrange multiplier. The extra equation required for complete solution is given by the constraint $G = 0$.

Alternatively, differentiating H with respect to λ and setting the result to zero yields the constraint and provides the same equation.

Consider the simple example of minimising the quadratic form

$$F(x, y) = x^2 + xy + y^2$$

subject to the constraint

$$G(x, y) = x + y - 1 = 0.$$

We form the function

$$H(x, y) = F + \lambda G = x^2 + y^2 + \lambda(x + y - 1)$$

then differentiate and set the result to zero

$$2x + y + \lambda = 0 \tag{A6.2}$$

$$x + 2y + \lambda = 0 \tag{A6.3}$$

$$x + y - 1 = 0. \tag{A6.4}$$

These two equations and the constraint give three simple simultaneous equations for the three unknowns x, y, λ, with solution $x = y = 1/2$, $\lambda = -3/2$. This example is simple enough to solve by substituting $y = 1 - x$ into F and differentiating with respect to x to give the same solution.

Minimising any quadratic function subject to a linear constraint always gives a set of simple simultaneous equations like (A6.2)–(A6.4), which makes the method particularly appropriate for linear inverse problems. The method of Lagrange multipliers is often the only possible approach when the constraint or optimising criterion is an integral rather than a simple function. More than one constraint can be introduced by using more Lagrange multipliers, one for each constraint.

A6.2 The penalty method

The penalty method is rather similar to the method of undetermined multipliers and is particularly simple to apply in a numerical solution. Instead of using an unknown multiplier λ, which must then be eliminated from the equations or solved for, we simply choose a large value for it. Minimising the combination H will then result in making G very small, because the contribution λG dominates. There remain enough degrees of freedom to also reduce F. F and G must be of a form that the ratio F/G cannot be made large by extreme values of the independent variables.

It remains to choose a suitable value for the *penalty parameter* λ, which is a numerical consideration. If λ is too large the optimising criterion will fall below numerical precision: only the constraint will be satisfied and the result will not achieve a minimum of F. When λ is too small the constraint will not be satisfied. The optimum value of λ depends on the acceptable ratio of F/G and machine precision; it may have to be found by trial and error.

Consider the simple example given above. Suppose we fix λ at some large value and use only (A6.2) and (A6.3) to find x and y. Equations (A6.2) and (A6.3) give $x = y = -\lambda/3$: large values of x and y therefore always result from large values of λ. This is because F is a quadratic function of x and y whereas G is a linear function, and F/G is maximised for $x, y \to \pm\infty$. This is not the solution we want. The problem is fixed by replacing $G = x + y - 1$ with its square, so that F/G remains finite as $x, y \to \pm\infty$. We now minimise $x^2 + xy + y^2 + \lambda(x + y - 1)^2$ to give

$$2x + y + 2\lambda(x + y - 1) = 0 \tag{A6.5}$$

$$x + 2y + 2\lambda(x + y - 1) = 0. \tag{A6.6}$$

Eliminating λ gives $x = y$. Equation (A6.5) then gives $x = 2\lambda/(3 + 4\lambda)$. As $\lambda \to \infty$ so $x, y \to 1/2$, as required. Choosing $\lambda = 100$ gives $x, y = 0.4963$ and $\lambda = 10^6$ gives $x, y = 0.499\,999\,6$. We could try to improve the solution by increasing λ still further, but we are limited by numerical accuracy. Suppose we retain six significant figures. The original equations (A6.5) and (A6.6) are completely dominated by the terms in λ; the other terms, which originated with the quantity we wish to minimise, are lost in the numerical error. The two equations become the same, except for round-off error, and solution would be impossible. In a larger problem the numerical difficulty would be more obscure to diagnose.

Appendix 7

Files for the computer exercises

A7.1 The practicals in this book

The computer practicals need publicly available software, as described here. Programs and data files specific to these practicals are available on the website `http://publishing.cambridge.org/resources/0521819652`. They were developed and run on a linux operating system and should run on any unix system with a FORTRAN compiler.

A7.2 The `pitsa` program

The `pitsa` program is used for the time series practicals in Part I. It was written by Frank Scherbaum and, at the time of writing, is available on the web at `http://lbutler.geo.uni-potsdam.de/service.htm`. Scherbaum's (1996) book *Of Poles and Zeros* makes extensive use of `pitsa`. Both program and book have been written for seismologists, but both are applicable to other subjects.

A7.3 `octave`

`octave` is now supported by the Free Software Foundation and is freely available from `www.octave.org`. It comes packaged with most linux systems. It closely resembles the commercial package `Matlab`, which the student may use instead if preferred because there are virtually no differences (apart from graphics) at the simple level used in this book.

A7.4 `gnuplot`

`gnuplot` is also supported by the Free Software Foundation and is available from `www.gnuplot.info`

A7.5 Other files

At the time of writing, other files used in this book may be freely downloaded from the website http://publishing.cambridge.org/resources/0521819652.

A7.6 Multi-taper

See Gabi Laske's website: http://mahi.ucsd.edu/Gabi/polaris.html.

References

Aki, K., Christoffersson, A. and Husebye, E. S. (1977). Determination of the three-dimensional structure of the lithosphere. *J. Geophys. Res.*, **82**, 277–96.

Ansell, J. H. and Smith, E. C. (1975). Detailed structure of a mantle seismic zone using the homogeneous station method. *Nature*, **253**, 518–20.

Arfken, G. B. and Weber, H. J. (2001). *Mathematical Methods for Physicists*. San Diego, CA: Harcourt Academic.

Backus, G., Parker, R. and Constable, C. (1996). *Foundations of Geomagnetism*. Cambridge: Cambridge University Press.

Backus, G. E. (1968). Kinematics of the secular variation. *Phil. Trans. R. Soc. Lond.* A **263**, 239–66.

Backus, G. E. (1970a). Non-uniqueness of the external geomagnetic field determined by surface intensity measurements. *J. Geophys. Res.*, **75**, 6339–41.

Backus, G. E. (1970b). Inference from inadequate and inaccurate data. i. *Proc. Nat. Acad. Sci. U.S.*, **65**, 1–7.

Backus, G. E. (1970c). Inference from inadequate and inaccurate data. ii. *Proc. Nat. Acad. Sci. U.S.*, **65**, 281–7.

Backus, G. E. (1970d). Inference from inadequate and inaccurate data. iii. *Proc. Nat. Acad. Sci. U.S.*, **67**, 282–9.

Backus, G. E. (1988). Bayesian inference in geomagnetism. *Geophys. J.*, **92**, 125–42.

Backus, G. E. and Gilbert, F. (1968). The resolving power of gross earth data. *Geophys. J. R. Astron. Soc.*, **16**, 169–205.

Bard, Y. (1974). *Non-linear Parameter Estimation*. New York: Academic Press.

Ben-Israel, A. and Greville, T. N. E. (1974). *Generalised Inverses: Theory and Applications*. New York: John Wiley.

Bloxham, J. (1986). Models of the magnetic field at the core–mantle boundary. *J. Geophys. Res.*, **91**, 13 954–66.

Bloxham, J., Gubbins, D. and Jackson, A. (1989). Geomagnetic secular variation. *Phil. Trans. R. Soc. Lond.*, A **329**, 415–502.

Box, G. E. P. and Tiao, G. C. (1973). *Bayesian Inference in Statistical Analysis*. London: Addison-Wesley.

Bullen, K. E. (1975). *The Earth's Density*. New York, NY: John Wiley & Sons.

Bullen, K. E. and Bolt, B. A. (1985). *An Introduction to the Theory of Seismology*. Cambridge: Cambridge University Press.

Christoffersson, A. and Husebye, E. S. (1979). On 3-dimensional inversion of P-wave travel time residuals. Option for geological modelling. *J. Geophys. Res.*, **84**, 6168–76.

Claerbout, J. F. (1992). *Earth Soundings Analysis: Processing Versus Inversion.* Oxford: Blackwell Scientific.

Constable, S. C., Heinson, G. S., Anderson, G. and White, A. (1997). Seafloor electromagnetic measurements above Axial Seamount, Juan de Fuca Ridge. *J. Geomagn. Geoelectr.*, **49**, 1327–42.

Cooley, J. W. and Tukey, J. W. (1965). An algorithm for the machine calculation of complex Fourier series. *Math. Computation*, **19**, 297–301.

Dahlen, F. A. (1982). The effect of data windows on the estimation of free oscillation parameters. *Geophys. J. R. Astron. Soc.*, **69**, 537–49.

Douglas, A. (1967). Joint epicentre determination. *Nature*, **215**, 47–8.

Ducruix, J., Courtillot, V. and Mouël, J. L. L. (1980). The late 1960s secular variation impulse, the eleven year magnetic variation and the electrical conductivity of the deep mantle. *Geophys. J. R. Astron. Soc.*, **61**, 73–94.

Duncan, P. M. and Garland, G. D. (1977). A gravity study of the Saguenay area, Quebec. *Canadian J. Earth Sci.*, **14**, 145–52.

Dziewonski, A. M. (1984). Mapping the lower mantle: determination of lateral heterogeneity in P velocity up to degree and order 6. *J. Geophys. Res.*, **89**, 5929–52.

Dziewonski, A. M., Bloch, S. and Landisman, M. (1969). A technique for the analysis of transient seismic signals. *Bull. Seism. Soc. Am.*, **59**, 427–44.

Eberhart-Phillips, D. and Reyners, M. (1999). Plate interface properties in the northeast Hikurangi subduction zone, New Zealand, from converted seismic waves. *Geophys. Res. Lett.*, **26**, 2565–8.

Franklin, J. N. (1970). Well-posed stochastic inversion of ill-posed problems. *J. Math. Annal. Appl.*, **31**, 682–716.

Garmany, J. D. (1979). On the inversion of travel times. *Geophys. Res. Lett.*, **6**, 277.

Gel'fand, I. M. and Levitan, B. M. (1955). On the determination of a differential equation by its spectral function. *Amer. Math. Soc. Transl.*, **1**, 253–304.

Gersztenkorn, A., Bednar, J. B. and Lines, L. R. (1986). Robust iterative inversion for the one-dimensional acoustic wave equation. *Geophysics*, **51**, 357–68.

Grand, S. P. (1987). Tomographic inversion for shear wave velocity beneath the North American plate. *J. Geophys. Res.*, **92**, 14 065–90.

Gubbins, D. (1975). Observational constraints in the generation process of the Earth's magnetic field. *Geophys. J. R. Astron. Soc.*, **47**, 19–31.

Gubbins, D. (1986). Global models of the magnetic field in historical times: Augmenting declination observations with archeo- and paleo-magnetic data. *J. Geomagn. Geoelectr.*, **38**, 715–20.

Gubbins, D. (1990). *Seismology and Plate Tectonics.* Cambridge: Cambridge University Press.

Gubbins, D. and Roberts, N. (1983). Use of the frozen-flux approximation in the interpretation of archeomagnetic and paleomagnetic data. *Geophys. J. R. Astron. Soc.*, **73**, 675–87.

Hanssen, A. (1997). Multidimensional multitaper spectral estimation. *Signal Processing*, **58**, 327–32.

Harris, F. J. (1978). On the use of windows for harmonic analysis with the discrete Fourier transform. *Proc. I.E.E.E.*, **66**, 51–83.

Hutcheson, K. and Gubbins, D. (1990). A model of the geomagnetic field for the 17th century. *J. Geophys. Res.*, **95**, 10 769–81.

Jackson, A., Jonkers, A. R. T. and Walker, M. R. (2000). Four centuries of geomagnetic secular variation from historical records. *Phil. Trans. R. Soc. Lond.*, A **358**, 957–90.

Jackson, D. D. (1979). The use of *a priori* data to resolve non-uniqueness in linear inversion. *Geophys. J. R. Astron. Soc.*, **57**, 137–57.

Jacobs, J. A. (1994). *Reversals of the Earth's Magnetic Field.* Cambridge: Cambridge University Press.

Johnson, L. E. and Gilbert, F. (1972). A new datum for use in the body wave travel time inverse problem. *Geophys. J. R. Astron. Soc.*, **30**, 373–80.

Kappus, M. E. and Vernon, F. L. (1991). Acoustic signature of thunder from seismic records. *J. Geophys. Res.*, **96**, 10 989–11 006.

Kellogg, O. D. (1953). *Foundations of Potential Theory.* New York: Dover Publishing Inc.

Khokhlov, A., Hulot, G. and Carlut, J. (2001). Towards a self-consistent approach to paleomagnetic field modelling. *Geophys. J. Int.*, **145**, 157–71.

King, N. E. and Agnew, D. C. (1991). How large is the retrograde annual wobble? *Geophys. Res. Lett.*, **18**, 1735–8.

Lall, U. and Mann, M. (1995). The Great Salt Lake – a barometer of low-frequency climatic variability. *Water Resources Research*, **31**, 2503–15.

Lay, T. and Wallace, T. C. (1995). *Modern Global Seismology.* London: Academic Press.

Lighthill, M. J. (1996). *An Introduction to Fourier Analysis and Generalised Functions.* Cambridge: Cambridge University Press.

Mao, W. J. and Gubbins, D. (1995). Simultaneous determination of time delays and stacking weights in seismic array beamforming. *Geophysics*, **80**, 491–502.

Mao, W. J. and Stuart, G. W. (1997). Transmission–reflection tomography: application to reverse VSP data. *Geophysics*, **62**, 884–94.

Masters, G., Johnson, S., Laske, G. and Bolton, H. (1996). A shear-velocity model of the mantle. *Phil. Trans. R. Soc. Lond.*, A **354**, 1385–411.

Michelini, A. and McEvilly, T. V. (1991). Seismological studies at Parkfield: I. simultaneous inversion for velocity structure and hypocenters using cubic b-splines parameterization. *Bull. Seism. Soc. Am.*, **81**, 524–52.

Neuberg, J. and Luckett, R. (1996). Seismo-volcanic sources on Stromboli volcano. *Ann. Geofisica*, **34**, 377–91.

Oldenburg, D. W. (1976). Calculation of Fourier Transforms by the Backus–Gilbert method. *Geophys. J. R. Astr. Soc.*, **44**, 413–31.

Papoulis, A. (1962). *The Fourier Integral and its Application.* McGraw Hill.

Pardoiguzquiza, E., Chicaolmo, M. and Rodrigueztovar, F. J. (1994). Cystrati – a computer program for spectral analysis of stratigraphic successions. *Computers and Geosciences*, **20**, 511–84.

Park, J., Lindberg, C. R. and Vernon, R. L. (1987). Multitaper spectral analysis of high-frequency seismograms. *J. Geophys. Res.*, **92**, 12 675–84.

Parker, R. L. (1977). Linear inference and underparameterized models. *Rev. Geophys. Space Phys.*, **15**, 446–56.

Pavlis, G. and Booker, J. (2000). Progressive multiple event location (PMEL). *Bull. Seism. Soc. Am.*, **73**, 1753–77.

Percival, D. B. and Walden, A. T. (1998). *Spectral Analysis for Physical Applications.* Cambridge: Cambridge University Press.

Press, W. H., Flannery, B. P., Teukolsky, S. A. and Vetterling, W. T. (1992). *Numerical Recipes.* Cambridge: Cambridge University Press.

Priestley, M. B. (1992). *Spectral Analysis and Time Series.* London: Academic Press.

Proctor, M. R. E. and Gubbins, D. (1990). Analysis of geomagnetic directional data. *Geophys. J. Int.*, **100**, 69–77.

Pujol, J. (1988). Comments on the joint determination of hypocentres and station corrections. *Bull. Seism. Soc. Am.*, **78**, 1179–89.

Robinson, E. A. and Treitel, S. (1980). *Geophysical Signal Analysis*. London: Prentice-Hall.

Sabatier, P. C. (1979). Comment on "The use of *a priori* data to resolve non-uniqueness in linear inversions" by D. D. Jackson. *Geophys. J. R. Astron. Soc.*, **58**, 523–4.

Sambridge, M. and Gallagher, K. (1993). Earthquake hypocenter location using genetic algorithms. *Bull. Seismol. Soc. Amer.*, **83**, 1467–91.

Scherbaum, F. (1996). *Of Poles and Zeros. Fundamentals of Digital Seismology*. Dordrecht: Kluwer Academic.

Schultz, M. H. (1973). *Spline Analysis*. Englewood Cliffs, New Jersey: Prentice-Hall.

Shearer, P. M. (1999). *Introduction to Seismology*. Cambridge: Cambridge University Press.

Spencer, C. (1985). The use of partitioned matrices in geophysical inversion problems. *Geophys. J. R. Astron. Soc.*, **80**, 619–29.

Stuart, G., Francis, D., Gubbins, D. and Smith, G. P. (1994). Tararua Broadband Array, North Island, New Zealand. *Bull. Seism. Soc. Am.*, **85**, 325–33.

Tarantola, A. (1987). *Inverse Problem Theory: Methods for Data Fitting and Model Parameter Estimation*. Amsterdam: Elsevier.

Thomson, D. J. (1982). Spectrum estimation and harmonic analysis. *Proc. IEEE*, **70**, 1055–96.

van der Hilst, R., Engdahl, E. R., Spakman, W. and Nolet, G. (1991). Tomographic imaging of subducted lithosphere below northwest Pacific Island arcs. *Nature*, **353**, 37–43.

vanDecar, J. C. and Crosson, R. S. (1990). Determination of teleseismic relative phase arrival times using multi-channel cross-correlation and least-squares. *Bull. Seismol. Soc. Amer.*, **80**, 150–69.

Walden, A. T. (1991). Some advances in nonparametric multiple time series and spectral analysis. *Econometrics*, **5**, 281–95.

Waldhauser, F. and Ellsworth, W. L. (2000). A double-difference earthquake location algorithm: Method and application to the northern Hayward fault, California. *Bull. Seismol. Soc. Amer.*, **90**, 1353–68.

Walker, M. R. and Jackson, A. (2000). Robust modelling of the Earth's magnetic field. *Geophys. J. Int.*, **143**, 799–808.

Whaler, K. A. (1980). Does the whole of the Earth's core convect? *Nature*, **287**, 528.

Yilmaz, O. (2001). *Seismic Data Analysis: Processing, Inversion, and Interpretation of Seismic Data*. Tulsa, OK: Society of Exploration Geophysicists.

Index

2D Fourier transform, *see* two-dimensional Fourier transform

a priori
  bound, 151
  information, 118
acausal filter, 60
ACH inversion, 181
Adams–Williamson equation, 112
aftershock, 58
airgun, 51, 170
Alfven's theorem, 209
aliasing, 40
  and null space, 160
  in 2D, 74
  tiling diagram, 74
amplitude, 34
  spectrum, 24, 162
angular frequency, 213
annihilator, 230
anti-alias filter, 42
array
  processing, 4, 170
  seismic, 170
associated Legendre function, 194
autocorrelation, 35, 63, 162
  DFT of, 35, 64
  symmetry, 162
averaging function, 122, 140
  designing, 144, 145
  on sphere, 203, 205
  width, 142
azimuth, 170

Backus ambiguity, 192
Backus effect, 205
Backus–Gilbert inversion, 142
bandpass filter, 57
bandwidth retention factor, 165
basis
  functions, 237
  vectors, 235
Bayesian inversion, 196
beam, 171
beamforming, 170–7
best fit
  global, 133
  local, 133
boxcar, 24
  taper, 164
  window, 225
Butterworth filter, 58
  in two dimensions, 75

causal filter, 60
checkerboard test, 182
completeness, 214
complex
  matrices, 228
  vector, 228
  vector space, 236
components of a vector, 235
confidence ellipsoid, 100
conjugate-gradient, 135
continuation
  downward, 189
  upward, 76
continuous inversion, 138–54
  discretising, 147
  recipe for, 153–4
convergence
  and damping, 127, 130
  criterion, 131
  nonuniform, 217
  radius of, 68
converted phase, 176
convolution, 161
  cyclic, 29–30
  theorem, 28–9, 34
    in 2D, 74
  with a spike, 34
Cook, James, 196
coordinate axes, 226
coordinates, 235

core
  Earth's, 87
  flow, 207–12
correlation, 97
  coefficient, 63, 94
covariance, 35, 131, 203
  of mapped errors, 118
  of model, 186
covariance matrix, 35, 97
  linearised, 129
  model, 100
cross correlation, 63, 171
cut-off frequencies, 57
cycle skipping, 134, 172
cyclic convolution, 29–30

damped inverse, 232
damping, 115–17, 172, 193–6
  and water level method, 163
  constant, choice of, 116
  core flow, 209
  step-length, 130
  undesirable effects, 211
data
  artificial, 119
  consistent with model, 134
  errors, 122
  kernel, 86, 138, 193
  rejection, method of, 103
  vector, 85
DC shift, 34
deconvolution, 10, 20, 65–9
  stabilising, 66
degrees of freedom, 122
delay, minimum, 61, 162
delay-and-sum, 171
delta function, 220–3
  derivative, 222
  Fourier integral transform of, 220
  substitution property, 221
density, 8
  Earth models, 112
deterministic, 10
detrend a sequence, 49
differentiation, 34
  of 2D data, 74
  of Fourier series, 30–1
digitise a time series, 49
dimension of vector space, 234
dip, 170
dip poles, 192
Dirac comb, 222
Dirac delta function, 140, *see* delta function
Dirichlet condition, 140
discrete
  convolution, 18
  convolution theorem, 18
  formulations, 10
  Fourier transform, 17, 21, *see* discrete Fourier transform
    of a spike, 24
    of autocorrelation, 35
    properties, 34
discrete Fourier transform
  2D, 73
  and filtering, 159
  periodicity of, 40
  time samples not equally spaced, 160
discretisation
  global, 148
  local, 148
  of a continuous inverse problem, 147
  point-wise, 148
distribution
  double-exponential, 104, 201
  Gaussian, 97, 201
  Laplace, 104
divergence theorem, 210
double-exponential distribution, 201
downward continuation, 77, 189
dual basis, 150
dynamic range, 3

Earth
  mass, 86
  moment of inertia, 86
  radius, 86
earthquake location, 127–30
  and inversion for structure, 183–7
eigenvalue, 226
  equation, 226
  of covariance matrix, 98
  of differential operator, 10
  zero, 114
eigenvector, 226
  left, 228
  right, 229
electrical heating, 195
ellipsoid, 227
  confidence, 100
  error, 100
  extreme values, 101
energy-delay theorem, 63
equations of condition, 85
  matrix, 89
equi-determined, 86
error
  ellipsoid, 100
  law of, 103
  surface, 132–5
  underestimate, 134
even function, 216
existence, 9, 148
expectation operator, 34
expected value, 35
exponential distribution, 201

f–k plane, 77
fast Fourier transform (FFT), 32–3
Fermat's principle, 179, 184
Feuillée, 196
FFT, *see* fast Fourier transform

field-aligned electric currents, 199
filter
  acausal, 60
  anti-alias, 42
  bandpass, 57
  Butterworth, 58
  causal, 60
  high-pass, 57
  low-pass, 57
  matched, 64
  minimum delay, 61
  non-realisable, 60
  notch, 57
  two-pass, 60
  Wiener, 161
  zero-phase, 59
filtering, 6, 57, 161
  in 2D, 75–7
final model, damping, 132
finite element, 148
forward
  modelling, 8, 125
  problem, 138
forward transform, 34
Fourier
  analysis
    as an inverse problem, 159–69
  coefficients, 213
  integral transform, *see also* Fourier integral transform (FIT), 218–23
  series, 11, 213–18
    complex, 21, 215
    half period, 216
    of monochromatic sine wave, 27
  transform, 10
    2D, 73–80
    symmetry, 33
Fourier integral transform
  estimate using inverse theory, 160
  existence, 218
  of exponential, 219
  of Gaussian, 219
  of monochromatic wave, 221
Fourier series
  and function space, 237
frequency
  angular, 213
  Nyquist, 40, 224
  sampling, 3, 21, 165
frozen flux, 189
  labelling magnetic field lines, 209
function space, 235–6

Gauss, 188
  coefficients, *see* geomagnetic coefficients
Gaussian distribution, 97
generalised inverse, 230–3
  of a rectangular matrix, 112
genetic algorithms, 126
geodynamo, 203
geological modelling option, 120
geomagnetic
  coefficients, 190, 195
  external field, 199
  jerk, 95
geomagnetism, 188–212
geometric progression, 22
geostrophic flow, 209
Gibbs's phenomenon, 44, 217
global best fit, 133
gradient, of 2D data, 75
Gram matrix, 141, 143
gravity
  traverse, 8, 73
  variations, 7
grid search, 126, 172

Herglotz–Wiechert, 177
Hermitian
  conjugate, 159, 228
  differential operators, 240
  matrix, 166
high-pass filter, 57
Hilbert space, 236
homogeneous station method, 186, 187
hyperbola, 173
hypocentre parameters, 127

IGRF, 188
induction equation, 210
inner product, 236
instrument response, 20
integral Fourier transform, 11
integration, 34
  of Fourier series, 30–1
International Geomagnetic Reference Field, *see* IGRF
International Seismological Centre, 104
inverse transforms, 34
inversion, 7
IRLS, iteratively reweighted least squares, 105

JHD, joint hypocentre determination, 186

kernel
  data, 86, 138, 193
  parameter, 150
knee of trade-off curve, 199
Kolmogoroff factorisation of the spectrum, 68
Kronecker delta, 23, 214
  substitution property, 23

lag, 63
Lagrange multipliers, 241–3
Laplace distribution, 104
Laplace equation, 189
leaky integration, 68
least squares
  approximation by Fourier series, 216
  iterative reweighted, *see* IRLS
  method, 162
  solution, 90
    unweighted, 151

Legendre functions, associated, 194
Legendre polynomials, 148, 238
Levinson's algorithm, 163
likelihood function, 97
linear
  algebra, 226–33
  differential operators, 239
  error analysis, 127
  independence, 234
  operators, 239
  parameter estimation, recipe for, 106
  regime, 131
  response, 3
  transformation, 98
linearisation, 125, 126
linearity, *see* response, linear
local best fit, 133
low-pass filter, 57
low-velocity zones, 180

magnetic
  land surveys, 197
  observatories, 197
Magsat, 199
main magnetic field, 188
mantle, Earth's, 87
mass, Earth's, 86
master event method, 186
matched filter, 64
matrix
  inversion
    by eigenvector expansion, 231
  sparse, 148
maximum frequency, 34, 165
maximum likelihood solution, 104
mean of random variables, 96
measurements, scientific, 85
microseismic noise, 1, 52
minimum
  delay, 61, 162
  phase, 61
minimum norm
  solution, 112, 114, 145, 232
    linear combination of the data kernels, 145
    spectrum, 160
model, 8
  resolution, 186
  vector, 85
moment of inertia, 86
Monte Carlo, 8, 126
multipliers, 140
multitaper approach to spectral analysis, 167

network seismic, 170
Newton–Raphson iteration, 126
noise, 5
  microseismic, 1, 52
  white, 36
non-symmetric matrices, 228
nonlinear inverse problem, 125–35
  methods, 125–7
  recipe, 135
  strategy, 126
norm, 112, *see* scalar product
  $\mathcal{L}^p$, 105
  minimum, 145
normal equations, 90
NORSAR, 120, 170, 178
notch filter, 57
null space, 110, 160, 210, 230
  of core flow, 209
null vector, 110
null-flux curves, 209
Nyquist frequency, 40, 224
Nyquist's theorem, 23

odd function, 216
optimal tapers, 165
orthogonal, 235
  functions
    expansion in, 148
    sines and cosines, 213
  matrix, 226
orthonormal, 235
  functions, 148
outliers, 103
over-determined, 86

pad with zeroes, 49
parabola
  Fourier series of, 215
parameter estimation, 9, 85–107
  Backus, 149
  nonlinear , 127
parametrisation, 11
  basis functions, 184
Parseval's theorem, 31–2, 166
  2D, 74
partial derivatives matrix, 129, 172
partitioned matrix, 229
pass band, 57
penalty
  method, 119–20, 242
  parameter, 242
period
  of Fourier series, 213
  of function, 213
periodic
  function, 213
  repetition, 28, 34
    and null space, 160
periodicity
  of DFT, 40
  of Fourier series, 216
phase, 61
  spectrum, 24, 34
  speed, 77
planes, pair of, as quadratic form, 227
polynomials
  and function space, 238
  Legendre, 148, 238
positivity constraint, 111

potential
  data, 76
  theory, 191
power spectrum, 11, 24, 64
PREM, 141
prewhitening, 163
prior
  constraint, 195
  information, 126
  probability distribution, 126
processing, 7
projection operator, 232
proton magnetometer, 50, 197

quadratic form, 227
quasi-linearisation, 126, 130–1
quelling, 180

radius of Earth, 86
random, 10
  variable, 34
    independent, 94
ranking, 113, 115
realisable filter, 60
recorder seismic, 3
rectangular matrix, 229–30
  eigenvalues, 229
  least squares inverse, 231
recursion
  deconvolution by, 20
  formula, 162
reference trace (in beamforming), 172
Reftek, 4
residual, 103
  statics, 173
resolution, 203, 205
  matrix, 121–3, 131, 205
    efficient way to compute, 122
    trace, 122, 205
  of an instrument, 3
  of continuous inversion, 142
response
  instrument, 20
  linear, 3
ringing, 196
robust methods, 103–6
root mean square, 99
rotational methods
  for solving equations, 120

sampling
  frequency, 3, 21, 165
  interval, 3, 23
  rate, 165
sawtooth, 215
scalar product, 235
secular variation, 189, 207
seismic
  array, 170
  marine surveys, 170
  network, 170
  recorder, 3
  reflector, 7
seismic tomography, 12
sensitivity, 3
sequence
  random, 34–36
  stationary, 35
  stationary stochastic, 35
Shannon's sampling theorem, 43, 224–5
shift theorem, 27, 34
  in 2D, 74
side constraint, 241
side lobes, 44
signal, 5
  band-limited, 49
similarity transformation, 98
sinc function, 43
slowness, 171, 177
Snell's law, 177
spatial anomalies, 74
spectral leakage, 45, 164
  parameter, 165
spectrum
  Kolmogoroff factorisation, 68
  power, 11, 24, 64
  practicalities, 49
  white, 36
spherical harmonics, 148, 190
spike, 23
  and delta function, 220
  Fourier transform, 24
spline approximations, 148
spread, 142
  function, 183
square wave, 214
SRO, 2
stabilise, 130
  an inversion, 114
stacking, 170
standard
  deviation, 94, 103
  error, 203
starting model, 126, 127, 172
stationarity, 12
stationary sequence, 35
statistical moments, higher, 127
step function, derivative of, 222
stochastic
  inversion, 118
  process, 149
  stationary sequence, 35
straight lines in 2DDFT, 77
stratified flow, 209
symmetric matrix, 226

taper, 49
  Blackman–Harris, 164
  boxcar, 164
  cosine, 164
  Gaussian, 164
  optimal, 165

tapering, 46
tiling diagram, 74
time reversal, 27, 34
time sequence, 17
Toeplitz matrix, 161–3
tomography, 177–83
  cross-borehole, 178
  medical, 177
  reflection–transmission, 178
toroidal flow, 210
tracers
  for core flow, 209
trade-off
  choice of parameter, 122
  curve, 115–17
    in continuous inversion, 143
  parameter, 115
  resolution and leakage, 164
travel time inversion, *see* tomography
travelling waves, 77
truncation, 196
  of expansion, 193
  of spherical harmonic series, 211
two-dimensional Fourier transform, 73–80
two-pass filter, 60

uncertainty principle, 46–7
uncorrelated, 35
under-determined, 86
  problem, 110–23
    recipe, 123
    recipe for, 123
undetermined multipliers, *see* Lagrange multipliers
uniform reduction, method of, 104
uniqueness, 9, 148, 190
  of solutions of Laplace's equation, 192
unit delay operator, 18
univariant, 98
  transformation to, 98
upward continuation, 76

variance, 35, 203
  of sum of random variables, 96
variational principle, 166, 227
vector space, 234–240
  complex, 236
  Euclidean, 235
  linear, 234

water-level method, 66, 163
wavenumber, 73, 74
wavevector, 74
weight matrix, 115
weighted average, 98
weighting directional measurements, 197
white noise, 36
white spectrum, 36
Wiener filter, 161
windowing, 44
winnowing, 113, 115
WWSSN, 2

z-transform, 17–21
zero-phase filter, 59